Hodder Gibson

Scottish Examination Mater

STANDARD GRADE
BIOLOGY

Problem Solving
in Biology
SECOND EDITION

James Torrance

Hodder Gibson

A MEMBER OF THE HODDER HEADLINE GROUP

Orders: please contact Bookpoint Ltd, 130 Milton Park, Abingdon, Oxon OX14 4SB. Telephone: (44) 01235 827720
Fax: (44) 01235 400454 Lines are open from 9.00–6.00, Monday to Saturday, with a 24 hour message answering service.
You can also order through our website www.hoddereducation.co.uk

British Library Cataloguing in Publication Data
Problem solving in biology. - New ed. - (Standard Grade science)
   1. Biology 2. Biology - Problems, exercises, etc.
   I. Torrance, James
   574'.076

   ISBN-10: 0-340-66406-1
   ISBN-13: 978-0-340-66406-3

Published by Hodder Gibson, 2a Christie Street, Paisley PA1 1NB.
Tel: 0141 848 1609; Fax: 0141 889 6315; email: hoddergibson@hodder.co.uk
First published 1996
Impression number   11 10
Year   2005

Copyright © 1996 James Torrance, James Fullarton, Clare Marsh, James Simms and Caroline Stevenson.

Cover photo of breaching Bottle-nose dolphins from Taxi, Getty (ba19430)
Typeset by Multiplex Techniques, Orpington, Kent.
Printed in Great Britain for Hodder Gibson, 2a Christie Street, Paisley, PA1 1NB, Scotland, UK

Contents

Preface

This book consists of a comprehensive bank of exercises intended for use by pupils studying Standard Grade and GCSE Level Biology courses and required therefore to develop skills in problem solving.

It is the aim of the book to give pupils practice in overtaking the extended grade criteria: handling and processing information; evaluating procedures and information; and drawing conclusions and making predictions.

The material is differentiated with some problems pitched at General Level, some carrying extra, more difficult questions and some aimed solely at the most able pupils.

The items are arranged into 25 Chapters. This allows for maximum flexibility of use by providing ready-made end-of-topic tests, extensions to classwork and homework assignments.

In addition to problem solving material, all but two of the chapters contain exercises requiring terms to be matched with their meanings. These have been included to allow pupils to gradually compile a glossary of the biological terms essential to Standard Grade Biology.

The Biosphere

1   Investigating an ecosystem

1 Match the terms in list X with their descriptions in list Y.

list **X**	list **Y**
1) abiotic	**a)** natural biological unit made up of living and non-living parts
2) abundance	**b)** group of living organisms of one type
3) biotic	**c)** place where an organism lives
4) community	**d)** method used to sample an ecosystem at regular intervals along a straight line
5) ecosystem	**e)** describing a factor that relates directly to some aspect of living things in an ecosystem
6) habitat	**f)** describing a factor that relates directly to a non-living feature of an ecosystem
7) line transect	**g)** rectangular-shaped sampling unit of known area
8) population	**h)** measure of the extent to which an organism occurs in an ecosystem (e.g. rare, common, etc.)
9) quadrat	**i)** all of the populations of plants, animals and micro-organisms that live together in an ecosystem

2 The following graph shows the range of pH within which each of six soil animals is found to occur.

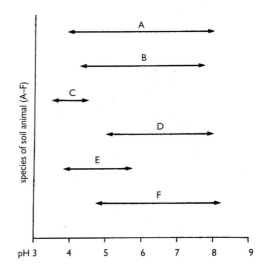

a) Which species occurs over the widest range of pH conditions?
b) Which species appears to be least tolerant of acidic conditions?
c) Which species is most tolerant of alkaline conditions?
d) Which species can survive in conditions of pH below 3.75?
e) How many species can tolerate pH 4?
f) Which of the following pH values can be tolerated by
(i) the largest number
(ii) the smallest number of species?
      3.5   4.5   5.5   6.5

3 The following table shows the number of different species of flowering plant and snail present in three areas of North America.

1

Problem Solving in Biology

area	latitude	flowering plants	snails
1	50° decreasing	650	30
2	40° distance	1625	91
3	30° from equator ↓	2100	172

a) What effect does latitude have on the number of different species of flowering plant and snail present?

b) Suggest a reason for this trend.

4 The following diagram shows a piece of ground viewed from above. It was sampled by taking four line transects (AB, CD, EF and GH) and recording the type of plant present at regular intervals along each transect.

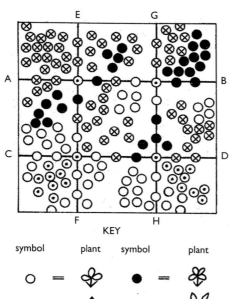

KEY

symbol	plant	symbol	plant
○ =	✿	● =	✾
⊗ =	⚘	⊙ =	⚘

a) Which of the four transects corresponds to the following diagram?

b) From the results of this one transect alone, which plant type seems to be most abundant? Give its symbol.

c) Look again at the diagram of the piece of ground. Is the plant that you gave as your answer to b) in fact the most abundant?

d) Describe a more accurate method of estimating which plant type is the most abundant in the area.

quadrats in randomly chosen sites
(each quadrat encloses one square metre)

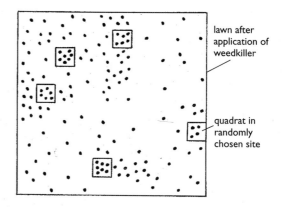

large lawn

dandelion plant

grass

5 The lawn in the above diagram measures 10 metres in length by 10 metres in breadth.

a) Calculate the area of the lawn.

b) Estimate the total number of dandelion plants growing on the lawn using only the information provided by the randomly chosen quadrats. The owner decided to try to remove the dandelions by spraying the lawn with selective weedkiller. The following diagram shows the lawn several weeks after spraying.

lawn after application of weedkiller

quadrat in randomly chosen site

c) Using the same method, estimate the total number of dandelion plants now growing on the lawn.

d) From a consideration of your results so far, does the weedkiller appear to have been effective? Explain your answer.

Taking an overall view of the lawn before and after spraying, the weedkiller does appear to have been at least partly successful.

e) (i) State the source of error in the sampling technique that was responsible for failing to show up this difference.

(ii) Explain how this error could be reduced to a minimum in a future investigation using the same sampling technique.

6 During an investigation of an ecosystem, pupils carried out tree-beating and caught the six animals shown in the following diagram. The second diagram shows an incomplete version of the branched key that they constructed.

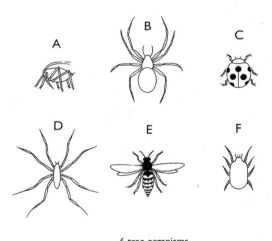

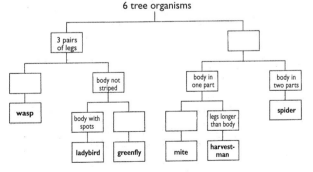

a) Copy and complete the branched key.
b) Use the key to identify animals A, B, C, D, E and F.

c) Convert the branched key into a key of paired statements.

7 The information in the diagram below and the key of paired statements that follows it refer to eight different fish.

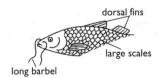
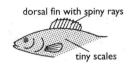

KEY

I	Body with large scales	..go to 2
	Body with tiny scales	..go to 5
2	One dorsal fin present	..go to 3
	Two dorsal fins present, the first with spiny rays	..go to 4
3	Dark spot behind head	**shad**
	Dorsal fin placed far back on body	**pike**
4	Dark spots on first dorsal fin forming bands	**zander**
	Dark spot on rear end of first dorsal fin	**perch**
5	Two or more barbels around the mouth	..go to 6
	No barbels present	..go to 7
6	One long barbel on lower jaw, two barbels at nostrils	**burbot**
	Two long barbels on upper lip	**catfish**
7	One long dorsal fin present	**blenny**
	Two dorsal fins present, the first with spiny rays	**miller's thumb**

a) Identify fishes W and X shown in following diagram.

b) Give TWO characteristics of a burbot.
c) Name TWO features shared by a zander and a perch.

d) Identify fishes Y and Z from the following descriptions: Fish Y has two dorsal fins, the first with spiny rays. It has a fairly broad head with the eyes on the top of it. It is a small fish and has neither barbels nor large scales. Fish Z lacks a lateral line but does have one black spot behind its head and one dorsal fin. It has large scales and those on its belly form a sharp toothed edge.
e) State TWO differences between a pike and a miller's thumb.

8 The graph below shows the effect of speed of water movement on the numbers of two species of insect larvae, A and B, found at the surface of a river.
a) At which range of water speed was the greatest number of species A recorded?
b) From which range of water speed was species A absent?

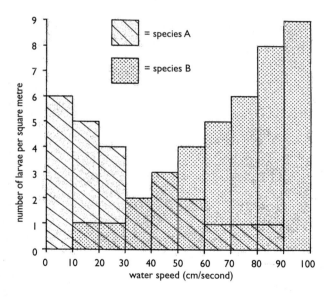

c) At which ranges of water speed were equal numbers of species A and B found?
d) For each species, state the relationship that exists between numbers present and the given ranges of water speed.

The following table shows the sizes of inanimate objects moved by different water speeds.

water speed (cm/second)	diameter of object (mm)	example of object
10	0.2	mud particle
25	1.3	sand particle
50	5.0	gravel particle
75	11.0	coarse gravel
100	20.0	pebble

e) Which species of insect larva would you expect to be more numerous in a stretch of river where the water speed is able to move particles of coarse gravel?

9 Two boys were asked to investigate the effect of light intensity on the distribution of meadow buttercup plants at the edge of the oak wood shown in the diagram opposite.

They pegged out a string line from point X outside the wood to point Y inside the wood. Along this line transect at points 1–10 they placed a metre square quadrat and counted the number of meadow buttercup plants present in each quadrat. This number was given an abundance score as shown in the accompanying table.

number of buttercup plants	abundance score	symbol
26 or more	abundant	◔
11–25	frequent	◑
6–10	occasional	◕
1–5	rare	◷
0	absent	○

The boys also measured the light intensity falling on each quadrat by taking one reading using a light meter, which gave readings on an 8-point scale of A–H where A = dimmest and H = brightest.

quadrat	1	2	3	4	5	6	7	8	9	10
light intensity	H	H	H	H	F	E	D	C	C	C
abundance of meadow buttercups	◕	◕	◕	◑	◑	◑	◔	◔	○	○

a) What abiotic factor was measured in the investigation?
b) In what way does the abiotic factor gradually change along the line transect from quadrat 1 to 10?
c) In what way does the abundance of buttercup plants change along the line transect from quadrat 1 to 10?
d) What relationship appears therefore to exist along the transect between abundance of buttercups and light intensity?
e) Suggest a possible reason for this apparent relationship.

f) Considering that some parts of the ground at the edge of a wood on a sunny day are lit up by patches of bright sunlight whereas others are in the shade, spot a shortcoming in technique used by the boys to measure light intensity.
g) Suggest how the shortcoming that you identified in question **f)** could be overcome.
h) It is possible that some other abiotic factor is wholly or partly responsible for the distribution of the buttercup plants. Suggest TWO such factors.

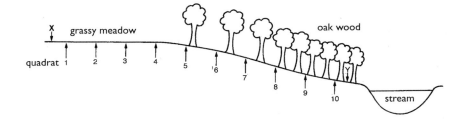

2 *How it works*

1 Match the terms in list X with their descriptions in list Y.

list X

1) birth rate
2) competition
3) consumer
4) death rate
5) food chain
6) food web
7) predator
8) prey
9) producer

list Y

a) green plant which makes food by photosynthesis
b) animal that hunts other animals for its food supply
c) relationship starting with a green plant followed by a series of animals each of which feeds on the previous one
d) struggle for existence between members of a community caused by the limited supply of an essential resource
e) general name for an organism unable to photosynthesise and dependent upon a ready-made food supply
f) animal that is hunted by other animals
g) measure of the number of new individuals produced by a population during a certain interval of time
h) complex relationship composed of several inter-related food chains
i) measure of the number of individuals within a population that died during a certain interval of time

2 Four types of plant typical of a woodland ecosystem are:
(i) trees that produce flowers and fruit (nuts) and shed their leaves in autumn;
(ii) mosses that have leaves but no flowers and grow at the base of trees;

5

Problem Solving in Biology

(iii) fungi that lack green leaves and feed on dead leaves from the trees;
(iv) small plants that grow under the trees and make their flowers in spring before the conditions become too shady.

a) Match each of these descriptions with the four plants given in the following chart.

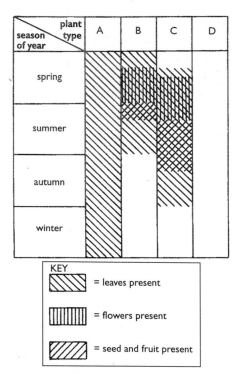

b) Which of these plants could provide a habitat and a food supply for a population of squirrels?

3 A 1-year-old carp fish has a mass of 30 g on average. A loch was stocked with a hundred of these young fish.
a) What was the total mass of this population when first put in the loch?
b) If a carp increases in mass by 50% each year, what was the total mass of the 100 fish after one year (assuming that they all survived)?
c) At the end of the first year 60% of the carp were caught by fishermen. What was the total mass of carp remaining in the loch?

4 The accompanying diagram shows the number of units of energy (in kilojoules/m²/year) that are transferred from organism to organism in a pond food chain.

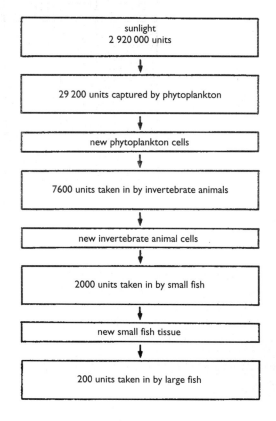

a) What percentage of the sunlight energy is successfully captured by the phytoplankton?
b) What percentage of energy is lost between intake of energy by small fish and intake of energy by large fish?

5 The graph shows the results from a study of the populations of two organisms, a predator and its well-fed prey, over a period of several weeks.
a) How many weeks did the population of prey organisms take to reach its maximum size?
b) How many weeks did the population of predators take to go from maximum numbers to minimum?

6

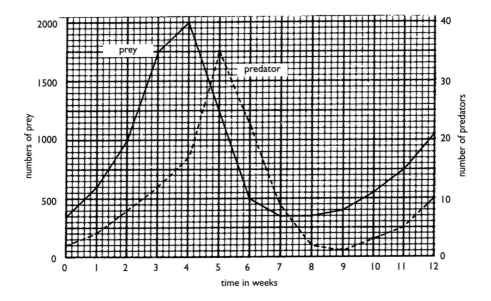

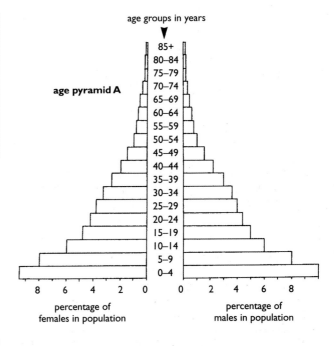

age groups in years

age pyramid A

c) By how many times was the total number of prey present at week 3 greater than that present at the start of the study?
d) What was the total number of predators present at week 6?
e) Account for (i) the decrease in prey and (ii) the increase in predators during the period between week 4 and week 5.

6 The following diagrams represent the population distributions for a developed country and a developing country.
a) In age pyramid A, (i) what percentage of the population are female aged between 5 and 9 years? (ii) Which age group makes up exactly 6% of the male population?
b) In age pyramid B, (i) what percentage of the female population are aged 65 and over? (ii) which sex has the larger percentage reaching old age?

c) Which age pyramid represents the population distribution of the developing country? Give TWO reasons for your answer.

Problem Solving in Biology

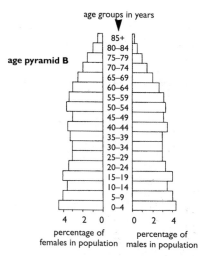

age groups in years

age pyramid B

percentage of females in population   percentage of males in population

7 The accompanying table refers to three different countries of the world.

a) Copy and complete the table.

	country		
	X	**Y**	**Z**
% birth rate	2	4	5
% death rate	1	2	
% overall population growth	1		2.5

b) Which of these countries is likely to have both the most effective birth control measures and the best medical services? Explain your answer.

8 The table below gives the results from a plant competition experiment where five groups of pea plants (each differing in number from the others) were grown in areas of similar fertile soil measuring ¼ square metre.

number of plants per ¼ square metre	average number of pods per plant	average number of seeds per pod
20	8.3	6.0
40	6.8	5.9
60	3.9	6.2
80	2.7	5.9
100	2.1	6.0

a) Plot the data in the table as two line graphs using graph paper similar to that in the diagram, which shows how to lay out the axes.

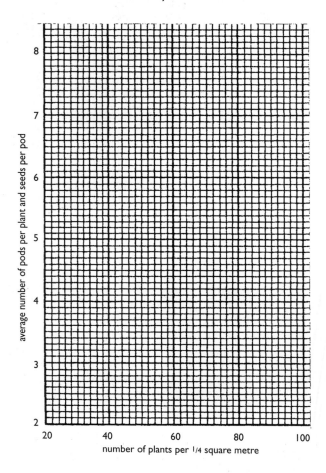

average number of pods per plant and seeds per pod

number of plants per ¼ square metre

b) Which feature appears to be unaffected by competition between neighbouring pea plants?

c) (i) Which feature appears to be affected by competition?

(ii) In what way is this feature affected as density of plants in a ¼ square metre increases?

(iii) Suggest TWO factors that neighbouring pea plants may be competing for.

d) Calculate the total number of seeds produced by
(i) 20 plants on a $^1/_4$ square metre;
(ii) 100 plants on a $^1/_4$ square metre;
e) Since seed mass is found to be unaffected by competition, which number of plants, 20 or 100, would be the better number to grow per $^1/_4$ square metre? Explain your answer.
f) It is possible that these results are unusual and not typical of pea plants in general. What should now be done to check the validity of these results?

9 The graph below shows the distribution of three species of organism at different depths in a loch where they form a food chain.

a) Which species has the widest vertical distribution in the loch?
b) Which species is the producer? Give TWO reasons to support your answer.
c) Identify the primary and the secondary consumers. Explain your choice in each case.

10 Imagine a sailor shipwrecked on a barren rocky island lacking top soil. From the ship's cargo he has managed to salvage one live hen and a bag of wheat grains.
a) To make these, his only resources, last for as long as possible, which of the following courses of action should he take?
A eat the hen and then eat the wheat
B feed all of the wheat to the hen, eat its eggs and then eat the hen when the wheat runs out
C share the wheat with the hen and then eat the hen when the wheat is finished
b) Justify your choice of answer.

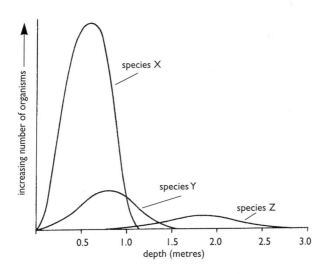

3   *Control and management*

I Match the terms in list X with their descriptions in list Y.

list **X**	list **Y**
1) fertiliser	**a)** contamination of environment by harmful substance
2) lichen	**b)** unwanted products of living things that are acted upon by decomposers
3) mayfly nymph	**c)** essential gas whose concentration is low in polluted water
4) organic waste	**d)** animal found in abundance in very badly polluted water
5) oxygen	**e)** chemical used to kill pests but pollutes food chains when used in excess
6) pesticide	**f)** poisonous gas that harms living things
7) pollution	**g)** simple plant composed of algae and fungus, which is especially sensitive to sulphur dioxide
8) rat-tailed maggot	**h)** animal found in abundance in oxygen-rich unpolluted water
9) sulphur dioxide	**i)** chemical used to improve soil but pollutes waterways when used in excess

Problem Solving in Biology

2 Repeated sampling of the water in the river shown in the diagram was done at sample points 1–5 during the years 1990 and 1994.

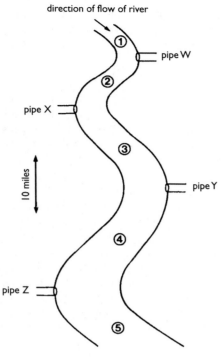

The results are summarised in the following table where:

■ = many   ▨ = few   ☐ = none

		1990					1994				
sampling point → indicator species		1	2	3	4	5	1	2	3	4	5
stonefly nymph			■			■	■	■	■	■	
caddis fly larva		▨	▨		■	▨	▨	▨	▨	▨	
bloodworm		▨	▨	■	■	▨	▨	▨	▨		■
rat-tailed maggot		▨	▨	■	▨	▨	▨	▨	▨		

a) (i) From which pipe was untreated sewage from an overloaded sewage works being discharged into the river in 1990? Explain your choice of answer.
(ii) Suggest a possible change that could have occurred to this sewage works during 1991–1993 to account for the result obtained in 1994.
b) In 1994 a paper mill began discharging untreated organic waste into the river through pipe Z.

Describe the effect this had on
(i) the type of animal most commonly found at sample site 5;
(ii) the oxygen concentration of the water at sample site 5.

3 The accompanying table shows the mass of sulphur dioxide produced by human activities and released into the atmosphere by an industrial European country over a period of 60 years.
a) Calculate the average mass of sulphur dioxide produced per year during the period shown.
b) Present the information as a line graph.
c) Draw TWO conclusions from the data.

year	mass of sulphur dioxide (million tonnes per year)
1920	3.8
1930	3.6
1940	4.4
1950	5.1
1960	6.3
1970	6.8
1980	5.0

4 Read the following passage and then answer the questions.

In the USA and many other countries, there is an enormous demand for cheap fast-foods such as beefburgers. The price of North American home-grown beef has rocketed in recent years, making it too expensive for use in low price burgers. Instead, cheap beef is imported from Central American countries such as Mexico. This beef comes from cattle raised on grasslands that used to be tropical forests.

It is to these same Central American forests (the ones that still remain) that North American songbirds such as the warbler migrate in autumn. However, as the forests are cleared, more and more songbirds fail to find a suitable overwinter habitat and therefore perish. Fewer songbirds return the next spring to North America where their food, countless species of young insect, are poised ready to devour crop plants. Until now predation by songbirds has kept the number of pests at a tolerable level but what of the future with songbird numbers declining by up to 4% per annum?

If Mexican ranchers could be persuaded to make better use of existing pasture lands, they would be able to produce about twice as much beef without clearing any more forests. But it is difficult to persuade landowners to invest money in fertilising and improving existing grasslands when new, temporarily rich, pastures can be quickly developed by clearing forested areas. However, in the long run this may prove to be a more realistic approach than attempting to persuade the average American fast-food devotee to turn vegetarian.

a) Why is Central American beef used for burgers instead of home-grown North American meat?
b) Construct a food chain from the information given in the first paragraph.
c) 'Removing one link in a food chain can lead to disaster'. Discuss the truth of this statement using an example contained in the passage.
d) (i) Give TWO possible solutions to the problem mentioned in paragraph 3.
(ii) Which ONE of these does the author suggest is less likely to be successful? Do you agree? Explain why.

5   The following information refers to a polluted river. Two readings have been omitted from the table.
a) Describe the relationship that exists between the number of green algae and the condition of the river water.

river organism	condition of river water				
	very clean	clean	fairly clean	dirty	very dirty
green algae	1	2	3	4	4
trout	3	1	0		0
water weeds	1	3		3	1

KEY TO ABUNDANCE LEVELS	
point on scale	description of population
0	absent
1	scarce
2	moderate
3	plentiful
4	abundant

b) Give the word that describes (i) the population of trout in clean water, (ii) the population of water weeds in dirty water.
c) From the choice given in brackets, select the appropriate number to indicate the most likely abundance of (i) trout in dirty water (0, 1, 2, 3), (ii) water weeds in fairly clean water (0, 1, 2, 3).

6   Read the following passage.

Crop yield by a piece of land is directly related to soil depth. Deeper soils produce better crops because they contain more water and mineral salts necessary for healthy plant growth.

Soil is a natural resource that renews itself only very slowly at a rate of 0.5 tonnes per hectare per year. However, annual soil loss in some parts of Britain is occurring at a rate of 30 tonnes per hectare per year. Soil that used to be 100 cm deep is now reduced to a thin layer of only 20 cm. This problem has become especially acute in recent years owing to farmers planting winter cereals that leave large areas of bare ground susceptible to soil erosion during the winter months. This is especially true of steeply sloping land that has been ploughed in a downslope line rather than along the contours (see diagram).

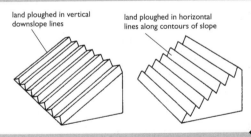

land ploughed in vertical downslope lines   land ploughed in horizontal lines along contours of slope

Instead of being offered financial incentives to conserve soil, farmers are encouraged by generous government subsidies to grow these high-yielding, autumn-planted winter cereals on any available land. This results in over-production of grain that then costs millions of pounds to store as a grain 'mountain' while the land that used to be over-wintered as grass continues to lose its top soil to wind and rain.

a) What natural resource is referred to in the above passage?

b) Explain why crop yield is directly related to soil depth.

c) By how many times is soil being lost at a faster rate than it is renewing itself in those parts of Britain referred to in paragraph 2?

d) (i) What name is given to this process of soil loss? (ii) Name TWO environmental factors that bring about this soil loss.

e) Give an example of a method of soil preparation carried out by some farmers that makes the problem of soil loss even worse.

f) Why do farmers plant winter cereals on every available hectare of soil?

g) Suggest a possible remedy to the problem of soil loss at (i) local level, (ii) national level.

7 The following map shows a region of coastline close to where a giant oil tanker was wrecked at sea. Prior to the disaster, the shallow waters of the coastline provided a rich source of edible crabs. Oil does not kill the crabs but harms their flesh, making them unsaleable. The extent of the shaded sector at each sample site represents the proportion of crabs with diseased flesh after the disaster.

a) Which sample site had the highest number of crabs?

b) In which sample site were the crabs only rarely found?

c) Describe the abundance level of crabs at sample site T.

d) Name the agent of pollution that affected the crabs.

e) In which sample site were fewest crabs affected by the pollutant?

f) Give two possible reasons for your answer to question (e).

g) In which sample site were most crabs affected relative to the population size present? Suggest why.

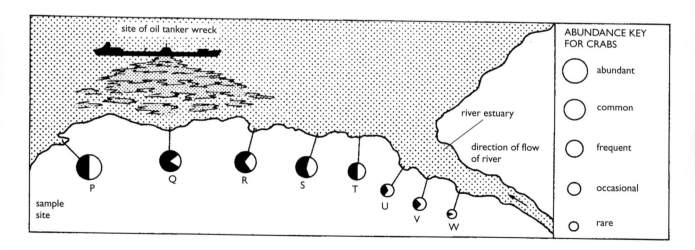

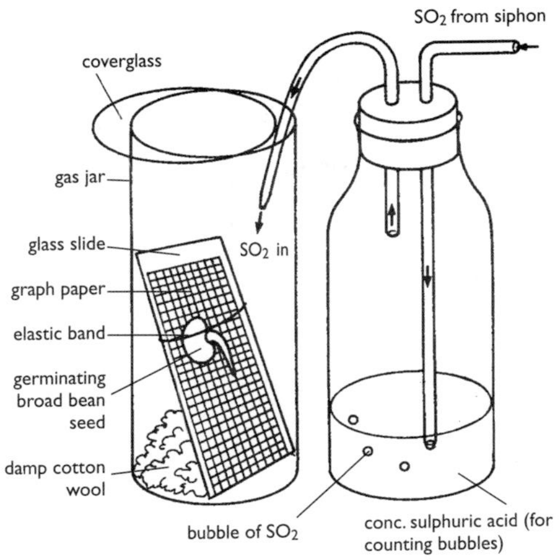

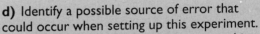

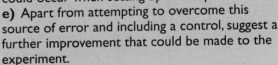

The increase in length of each young root was measured after 5 days and the results are shown in the following table:

number of SO₂ bubbles	increase in length of root after 5 days (mm)
5	6
10	2
20	0.5

a) Suggest the reason for using (i) the graph paper, (ii) the damp cotton wool.

b) What control should have been included in the experiment?

c) Predict the effect of using 30 bubbles of SO₂.

d) Identify a possible source of error that could occur when setting up this experiment.

e) Apart from attempting to overcome this source of error and including a control, suggest a further improvement that could be made to the experiment.

labels on diagram:

coverglass

gas jar

glass slide

graph paper

elastic band

germinating broad bean seed

damp cotton wool

SO₂ from siphon

SO₂ in

bubble of SO₂

conc. sulphuric acid (for counting bubbles)

8 The above apparatus shows a method used to investigate the effect of sulphur dioxide (SO₂) gas on the growth of young roots.

Three gas jars, each containing a germinating broad bean seed, were set up as shown. Using a SO₂ siphon, 5 bubbles of SO₂ were passed into the first gas jar. The second gas jar received 10 and the third jar 20 bubbles of SO₂. Each gas jar was sealed using a coverglass.

Investigating Cells

4 Investigating living cells

1 Match the terms in list X with their descriptions in list Y.

list X

1) cell
2) cell membrane
3) cell sap
4) cell wall
5) chloroplast
6) cytoplasm
7) nucleus
8) permanent vacuole
9) temporary vacuole

list Y

a) structure that controls cell's activities

b) green structure containing chlorophyll found in some plant cells

c) large centrally located cavity full of cell sap in plant cells

d) basic unit of which living things are composed

e) solution of sugar and salts

f) fairly rigid boundary round plant cell that maintains the cell's shape

g) small changeable cavity containing food or water found in some types of animal cell

h) transparent jelly-like material found in a cell, which is the site of biochemical reactions

i) thin flexible boundary surrounding a living cell, which controls entry and exit of materials

2 a) Steps i)–vi) listed below give the procedure adopted to use the microscope shown in the accompanying diagram. Arrange them into the correct order starting with iv).
(i) Look through P and bring the specimen into focus using Q and R.

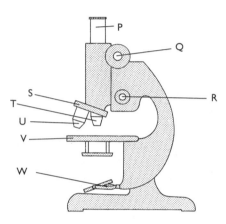

(ii) Arrange the prepared slide with the specimen above the centre of the hole in V.
(iii) Change to high power by turning S until U is above the specimen.
(iv) Check that T is exactly above the hole in V.
(v) Use Q to lower T to about 5 mm from the prepared slide.
(vi) Move W until light passes up through the microscope.

b) (i) A girl set up this type of microscope to view onion cells but she found that all of the image was very dark. Suggest the part of the procedure that she should check.
(ii) A boy found that one half of the image was brightly illuminated and the other half was in darkness. Suggest a remedy from the procedure list.
(iii) When looking at onion cells, a girl found that the image was spoiled by a black mark, which unlike the image of the cells, revolved when she turned the eyepiece. Suggest a solution to this problem (not listed in part **a)** above).

c) If P contains a lens with a magnification of x15 and T and U give a magnification of x8 and x40 respectively, which of the following is the highest magnification possible for this microscope?
A x55   **B** x120   **C** x600   **D** x4800

3 The branched key in the diagram below refers to ten cell types. Use only the information given in the key to answer the questions that follow it.

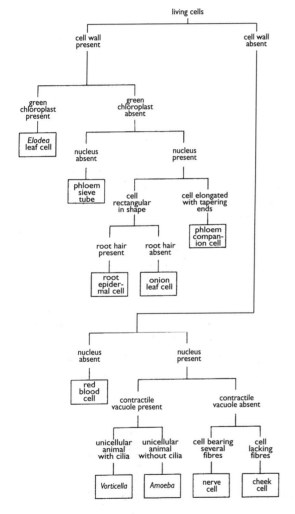

a) Identify cell types X and Y shown in the following diagram.
b) Give THREE structural features of *Vorticella*.
c) State ONE difference between a phloem sieve tube and an onion leaf cell.
d) Make a simple diagram of a phloem companion cell.

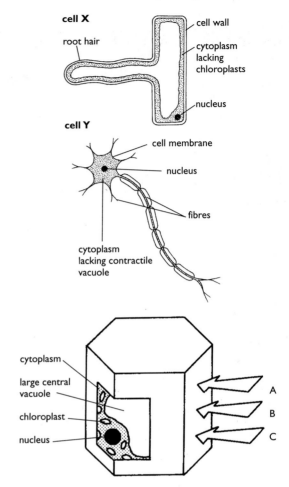

4 The diagram above shows a 3-dimensional view of a plant cell with a 'window' cut out to reveal some of its internal structures.
    Imagine that the cell has been sliced across at planes A, B and C. Match these with the diagrams of the cell's internal structure, 1, 2 and 3.

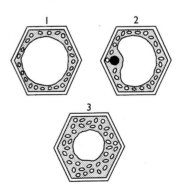

Problem Solving in Biology

5 The unit used to measure cell size is the micrometre. 1 millimetre = 1000 micrometres.
a) A human red blood cell is 7 micrometres in diameter. Express this as a decimal fraction of a millimetre.
b) If a cell is 0.008 millimetres long, what is its length in micrometres?

6 Present the information given in the following table as a bar chart.

cell type	cell length or diameter (in micrometres)
onion epidermis	150
human egg	100
sea urchin egg	70
Paramecium	50
human liver	20
yeast	8
human red blood corpuscle	7
Bacillus bacterium	3

7 Two purposes of brown iodine solution in a biological laboratory are: to stain structures (if a structure absorbs more iodine solution than its surroundings, it becomes a darker brown colour); and to test for starch (if starch is present, it turns blue–black in the presence of iodine solution).

The following diagram shows three types of plant cell before and after the addition of iodine solution.
a) (i) Which cellular structure became stained brown?
(ii) In which cell types was the process of staining observed in this experiment?
b) (i) Which cell type contains starch grains?
(ii) Give a reason for your answer.

	before	after
green cell from *Elodea* leaf		chloroplast, cell wall, nucleus (dark brown)
cell from potato tuber		blue–black structure, cytoplasm, cell wall
cell from onion epidermis		cell wall, cytoplasm, nucleus (dark brown)

8 Read the following passage.

If the nucleus of a cell, or some other structure such as a chloroplast, is removed from the cell, the isolated part cannot survive on its own. A complete cell is the smallest unit that can lead an independent life. This is neatly illustrated by unicellular *Amoeba*, a tiny animal that shows all the characteristics of living things despite the fact that it consists of only one cell.

Multicellular animals and plants are made of more than one cell. A human adult's body is composed of approximately 60 billion cells. Instead of each one of these cells performing every function vital for life, it is more efficient for certain cells to become specialised to do particular jobs.

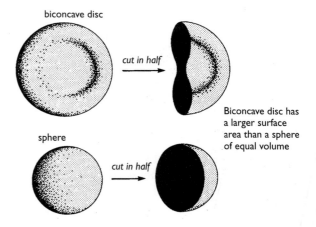

biconcave disc

cut in half

sphere

cut in half

Biconcave disc has a larger surface area than a sphere of equal volume

A red blood cell, for example, is shaped like a biconcave disc as shown in the diagram. This shape presents a large surface area through which oxygen can pass into the cell before being transported to all parts of the body. As a result of its shape, a red blood cell is very good at this job.

A group of similar cells working together and carrying out a particular function is called a tissue. For example, muscle tissue brings about movement and nerve tissue carries messages. A group of different tissues in turn make up an organ. For example the heart is made of different tissues such as muscle, connective and nerve tissues working together to pump blood around the human body.

a) Explain why a cell is often described as 'the basic unit of life'.
b) Line 7 refers to '...all the characteristics of living things ...'. List as many of these as you can.
c) Distinguish between unicellular and multicellular living things.

d) Another way of expressing the number 1000 is 10^3.
(i) A human baby is made up of about 2 billion cells. Express this number in the same way (where 1 billion = 1 million million).
(ii) By how many times is the number of cells in a human adult greater than that in a baby?
e) Why is a red blood cell a good example of a cell whose structure is ideally suited to the job that it does?
f) What general name is given to a group of similar cells working together to do a particular job?
g) Human skin is composed of many structures such as blood vessels, epidermal cells, nerve endings, sweat glands, etc. Is skin a tissue or an organ?
h) Arrange the following in order of increasing complexity: organ, cell, human body, tissue.

9 The average length of a certain type of cell was found to be 150 micrometres. How many of this type of cell would need to be laid end to end to form a line the same length as this page? (Answer to nearest whole cell.)

5 *Investigating diffusion*

I The diagram shows an experiment set up to investigate the process of diffusion. After ten minutes a distinct blue–black colour was found to be present inside the visking tubing sausage but not outside it.

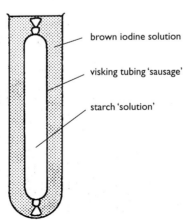

brown iodine solution

visking tubing 'sausage'

starch 'solution'

a) Rewrite the following sentence choosing the correct word.
Iodine solution has reacted with starch inside/outside the visking tubing 'sausage'.
b) It was concluded from the experiment that molecules of one of the two liquids were small enough to diffuse through the visking tubing membrane. Identify this liquid.
c) It was concluded that molecules of one of the liquids were too large to diffuse through the membrane. Identify this liquid.

d) An alternative explanation of the results is that the membrane allows molecules of any size to pass in but not out.
(i) Describe the experiment that you would set up to investigate this theory.
(ii) Explain how you would know from your results whether the theory is true or false.

Problem Solving in Biology

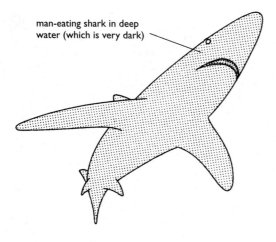

man-eating shark in deep water (which is very dark)

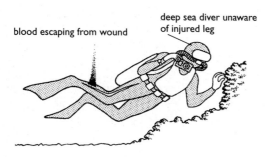

blood escaping from wound

deep sea diver unaware of injured leg

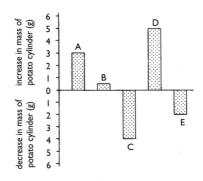

a) Which cylinder had been immersed in sugar solution of (i) the highest concentration (ii) the lowest concentration?
b) Predict what would have happened to the mass of a potato cylinder if it had been immersed in a sugar solution exactly equal in concentration to that of its cells.
c) Which of the five cylinders in the experiment had been immersed in sugar solution nearest in solution concentration to that of potato cells?
d) Calculate the percentage increase in mass shown by cylinder A.

2 Write a short paragraph to explain why the deep sea diver shown in the diagram is in danger. Use all the following words and phrases in your answer: low concentration, particles of blood, diffusion, high concentration, deep water.

3 Five identical cylinders of fresh potato (A–E), each weighing 15 g, were immersed in sugar solutions of different concentration for 3 hours and then dried and reweighed. The change in mass of each cylinder was recorded in a bar graph as shown in the accompanying diagram.

4 In the experiment shown below, the level of sugar solution was found to rise up the tube. The level was read at regular intervals and then results graphed as shown.
a) What evidence is there from this experiment that visking tubing is a selectively permeable membrane?
b) What was the height of the sugar level at minute 4?
c) How much longer did it take for the sugar level to reach the top of the tube?
d) Predict the effect of using a more dilute sugar solution on the time taken for the level to reach the top of the tube.

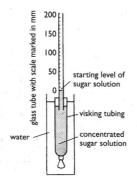

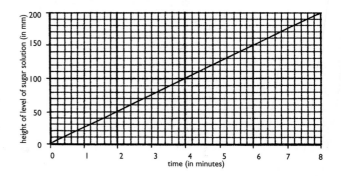

5 The following diagram shows an experiment set up to investigate the process of osmosis.

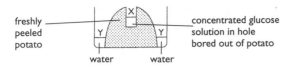

freshly peeled potato — X — concentrated glucose solution in hole bored out of potato

Y   Y

water   water

a) After a few hours, which of the following would have occurred?

A A rise in level X and a rise in level Y.

B A drop in level X and a drop in level Y.

C A rise in level X and a drop in level Y.

D A drop in level X and rise in level Y.

b) Which of the set-ups in the diagram below would be a suitable control for the above experiment?

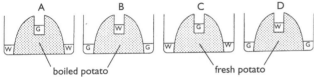

A   B   C   D

boiled potato   fresh potato

W = water
G = glucose solution

6 When sugar is sprinkled onto fresh strawberries, liquid is found to collect in the dish. Suggest a reason for this.

7 Match the terms in list X with their descriptions in list Y.

list X	list Y
1) concentration gradient	**a)** structure that allows rapid passage through it of water molecules, that are small, but not larger molecules
2) diffusion	**b)** region to which net movement of water molecules occurs during osmosis
3) freely permeable membrane	**c)** term used to describe a plant cell whose contents have shrunk due to loss of water by osmosis
4) higher water concentration	**d)** the difference in concentration that exists between two regions before diffusion occurs

5) lower water concentration	**e)** movement of molecules of a substance from high to low concentration
6) osmosis	**f)** term used to describe structure that allows rapid passage through it of all molecules in solution
7) plasmolysed	**g)** term used to describe plant cell swollen with water taken in by osmosis
8) selectively permeable membrane	**h)** process of increased movement of water molecules through a selectively permeable membrane to a more concentrated solution
9) turgid	**i)** region from which net movement of water molecules occurs during osmosis

8 A piece of rhubarb epidermis tissue was immersed for 5 minutes in each of three liquids (A, B and C), one after the other, and the cells examined at the end of each 5 minute period. The following diagram shows the appearance of one typical cell from the tissue after each immersion.

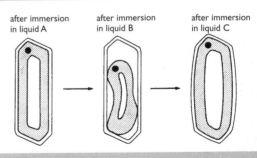

after immersion in liquid A   after immersion in liquid B   after immersion in liquid C

a) Match the three liquids with the following: water, dilute sugar solution, concentrated sugar solution.

b) Immersion in which liquid brought about plasmolysis?

9 A type of microscopic organism lives in the estuary of the river shown in the first diagram overleaf. Two high and two low tides occur daily and these affect the organism's volume as indicated by the graph in the second diagram.

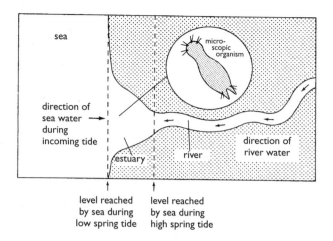

sea

micro-scopic organism

direction of sea water during incoming tide

direction of river water

estuary   river

level reached by sea during low spring tide

level reached by sea during high spring tide

(i) Compared with sea water, river water has a higher/lower salt concentration.
(ii) When the tide is low, the salt concentration in the river estuary will increase/decrease.
(iii) When the tide is high, the salt concentration in the river estuary will increase/decrease.
b) (i) Between which TWO sets of times of day was the animal gaining water by osmosis?
(ii) Was the tide coming in or going out during these times?
c) (i) Between which TWO sets of times of day was the animal losing water by osmosis?
(ii) Was the tide coming in or going out during these times?
d) The data in the graph refer to the average volume of 100 organisms. Why were so many used?

a) Rewrite the following sentences, choosing the correct word at each choice.

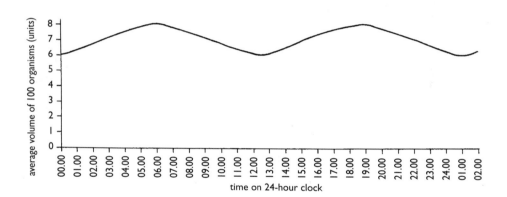

time on 24-hour clock

6 *Investigating cell division*

I   Match the terms in list X with their descriptions in list Y.

list X

1) centromere

2) chromatid

3) chromosome

list Y

a) nuclear division resulting in formation of two identical daughter nuclei

b) process following nuclear division that completes cell division

c) central region of cell where chromosomes are found immediately prior to nuclear division

4) cytoplasmic division

5) equator

6) identical daughter cells

7) interphase

8) mitosis

d) delicate structure connecting centromere of a chromosome to one of cell's poles

e) one of two identical replicas of a chromosome

f) threadlike structure carrying genetic code found inside nucleus of a living cell

g) cellular structure that divides during mitosis

h) structure that temporarily holds two identical chromatids together

9) nucleus

i) phase of cell development when each chromosome forms an exact replica of itself

10) spindle fibre

j) end products of mitosis each containing a complete set of chromosomes

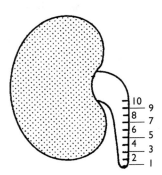

2 The above diagram shows a germinating broad bean seed. Its young root has been marked at 1mm intervals with waterproof ink.

It was allowed to grow for a further three days and was then carefully examined.

Mitosis had occurred between marks 1 and 2 only. The cells in the region marked 2–5 had not divided but instead had become elongated. The cells in the region marked 5–10 had neither divided nor elongated.

a) Which of the following diagrams best represents the young root three days after receiving its ink marks?

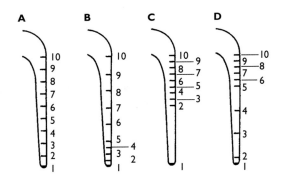

b) (i) The diagram below shows some cells at different stages of mitosis and cell division from between marks 1 and 2 in the first diagram. Arrange them into the correct sequence starting with W.

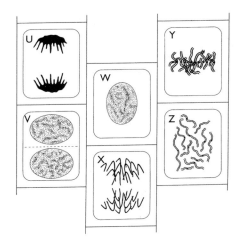

(ii) Strictly speaking, which of these are stages in the process of mitosis (nuclear division)?

(iii) Which ONE of choices **A, B, C** and **D** in the following table correctly matches three of the stages with the events taking place?

	event		
	attachment of chromosomes on equator	**separation of chromatids**	**division of cytoplasm**
A	Y	X	V
B	U	X	V
C	Y	Z	V
D	Y	X	U

3 Each normal body cell in a kangaroo contains 12 chromosomes. Which of the following diagrams correctly represents two successive cycles of mitosis and cell division? (The numbers refer to the chromosomes present in each cell.)

Problem Solving in Biology

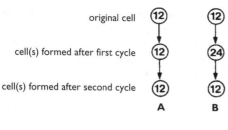

original cell

cell(s) formed after first cycle

cell(s) formed after second cycle

A   B

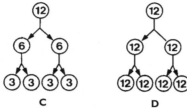

C   D

4 The bar chart in the accompanying diagram refers to the chromosome complement (total number of chromosomes in a cell) of seven crop plants.

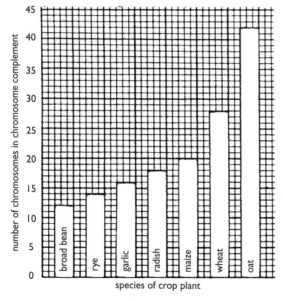

a) State the number of chromosomes present in the chromosome complement of a garlic plant.
b) Compared with a rye plant, how many more chromosomes are present in the complement of a wheat plant?
c) Express as a simple whole number ratio the chromosome complement of oat to rye.

d) In terms of chromosome number, the chromosome complement of oat exceeds that of one of the others by 3.5 times. Identify the plant.
e) Each chromosome consists of units of heredity called genes. Why would it be incorrect to conclude from the information in the bar chart that a radish plant definitely contains more genes in its chromosome complement than a garlic plant?

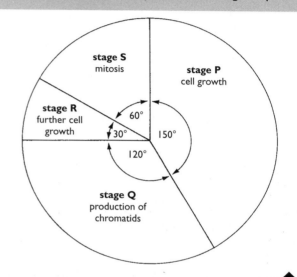

5 The stages that occur before, during and after mitosis in a certain type of animal cell are represented by the pie chart shown above. One complete cycle takes 12 hours at 22°C.

a) Copy and complete the following table.

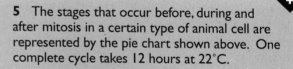

stage	time taken at 22°C (hours)
P	
Q	
R	
S	

b) (i) Express as a whole number ratio, the total time spent on cell growth to the time spent on production of duplicate chromosomes at 22°C.
(ii) Express the time spent on mitosis as a percentage of the total time required for one cycle at 22°C.

c) At 38°C, the cycle takes 4 hours but the relative time required for each stage remains unchanged. Copy the following sentences, choosing only the correct answer from each bracketed set of alternatives.

An increase in temperature $\left\{\begin{array}{l}\text{slows down}\\ \text{fails to affect}\\ \text{speeds up}\end{array}\right\}$ the rate of

mitosis and cell division. In this type of cell at 38°C,

stage Q would take $\left\{\begin{array}{l}\text{1 hour 20 minutes}\\ \text{1 hour 30 minutes}\\ \text{1 hour 40 minutes}\end{array}\right\}$

and stage S would take $\left\{\begin{array}{l}\text{30 minutes}\\ \text{40 minutes}\\ \text{60 minutes}\end{array}\right\}$

d) Predict the effect on the rate of mitosis and cell division of lowering the temperature to 10°C.

6 A student observed cells dividing under a microscope and recorded her results as shown in the table below.

The distance to which the table refers is the average distance between the centromeres of the chromosomes and the poles of the spindle.

time (minutes)	distance (micrometres)
0	20
5	20
10	9
15	2
20	0

a) Plot the data as a line graph.
b) During which time interval did the sister chromatids begin to move apart?
c) Explain how you arrived at your answer to **b)**.

7 The diagram below shows some of the stages in the development of a certain multicellular animal. A series of cell divisions of the fertilised egg in a vertical and horizontal plane results in the formation of an embryo.
a) Copy and complete the accompanying diagram.
b) Calculate the number of cells present in the embryo after three further cell divisions.

c) (i) State the relationship between cell number and cell size during embryo formation as shown by the diagram you drew as your answer to question **a)**.
(ii) Does this relationship also apply to the size of the nucleus? Explain your answer.

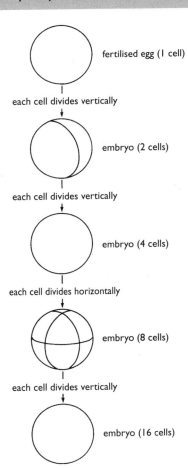

fertilised egg (1 cell)

each cell divides vertically

embryo (2 cells)

each cell divides vertically

embryo (4 cells)

each cell divides horizontally

embryo (8 cells)

each cell divides vertically

embryo (16 cells)

7 *Investigating enzymes*

I Match the terms in list X with their descriptions in list Y.

list X

1) catalyst

2) denatured

3) digestion

4) enzyme

5) product

6) protein

7) substrate

8) synthesis

list Y

a) building-up of large complex molecules from simpler ones by an enzyme-controlled reaction

b) type of organic chemical of which enzymes are composed

c) protein made by living cells that acts as a biological catalyst

d) chemical that increases the rate of a chemical reaction and remains unaltered

e) substance upon which an enzyme acts resulting in the formation of an end product

f) enzyme-controlled breakdown of large complex molecules to simpler ones

g) term used to describe the state of an enzyme that has been permanently destroyed

h) substance formed as a result of an enzyme acting on its substrate

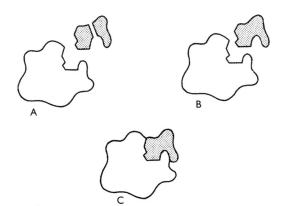

2 The above diagrams show the stages that occur during the action of an enzyme to promote a breakdown reaction. Using the three letters given and arrows, draw a flow diagram to indicate the correct sequence in which the three stages would occur.

3 The following table gives the results from an experiment involving the breakdown of a food by an enzyme.

temperature (°C)	mass of food broken down (mg/hour)
5	3
15	9
25	17
35	21
45	18
55	1

a) Using similar graph paper and the axes shown in the diagram (which have been only partly completed), present the result as a line graph.

b) State the temperature at which the enzyme is most active.

c) In your opinion, what does the graph suggest about the mass of food that would be broken down at 65°C?

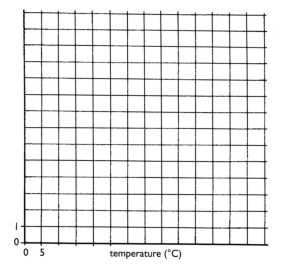

4 The experiment in the diagram opposite was set up to investigate the effect of the enzyme salivary amylase on starch 'solution'. Two equal lengths of visking tubing were filled with 1% starch 'solution', rinsed and placed in test tubes of water. Both test tubes were kept in the water bath at 37°C for 1 hour.

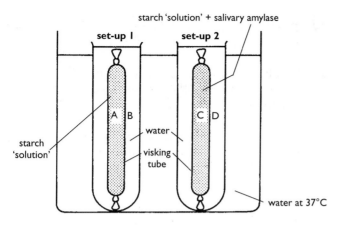

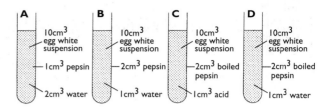

starch 'solution' + salivary amylase

a) Which ONE of the four designs is a valid control?
b) Explain, in turn, why each of the others is invalid.

Samples of liquid from regions A, B, C and D were tested for sugar and starch at the start and after one hour. The results are shown in the following table. (+ = food present, − = food absent)

	sugar test				starch test			
	A	B	C	D	A	B	C	D
at start	−	−	−	−	+	−	+	−
after 1 hour	−	−	+	+	+	−	−	−

a) Which set-up is the control?
b) Name THREE features of the experimental procedure that are essential for a valid comparison of set-ups 1 and 2 to be made.

c) Draw TWO conclusions from the results of the experiment.

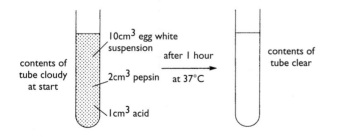

5 The experiment shown in the above diagram was carried out by a group of pupils to investigate the effect of acid on the action of the enzyme pepsin. Other groups were asked to design a control experiment. Some of their suggestions are shown in the following diagram.

6 Plant amylase is an enzyme that digests starch to simple reducing sugar. In an experiment, an equal volume of plant amylase was added to a hole in the centre of each of six petri dishes containing starch agar. Each plate was kept at a different temperature for 24 hours and then the surface of the starch agar was flooded with iodine solution. The diameter of the non−blue−black zone round each hole was measured as an indication of the amount of enzyme activity (since the starch in this zone had been digested).

The following diagram shows the results plotted on axes as six points.

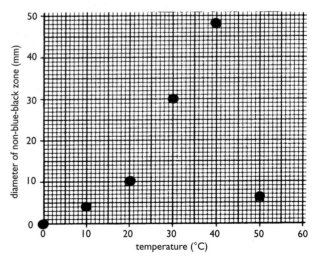

a) Using similar graph paper, copy and complete the graph.
b) Copy and complete the following table to show the data that gave rise to the six plotted points.

	diameter of non blue–black zone (mm)
0	
10	
	10
30	
	48
50	

c) Using your graph, state the diameter of the non–blue–black zone that would have resulted at (i) 25°C and (ii) 35°C.

d) Predict the diameter of the non–blue–black zone that would have been formed in a plate kept at 60°C. Explain your answer.

e) This experiment has been criticised because, strictly speaking, it contains a second variable factor that may affect the results. Identify this factor and suggest the general effect that it may have on the results.

7 One gram of roughly chopped liver was added to hydrogen peroxide solution at different pH values and the time taken to collect 1 cm³ of oxygen was noted in each case. The results are given in the following table.

pH of hydrogen peroxide solution	time to collect 1 cm³ of oxygen (seconds)
6	105
7	78
8	57
9	45
10	52
11	66
12	99

a) Present the results in the form of a line graph with pH on the horizontal axis.

b) From your graph state the pH at which the enzyme was (i) most active, (ii) least active.

c) Of the pH values used in this experiment, which is the optimum for the enzyme present in the liver cells?

d) How could you obtain an even more accurate measurement of the optimum pH at which this enzyme works?

8 The bar graph in the diagram shows the results from an investigation into the effect of alcohol on the activity of an enzyme that digests food in the small intestine of humans.

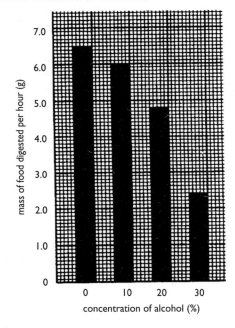

a) In general what effect does increasing concentration of alcohol have on the mass of food digested per hour?

b) Which concentration of alcohol has the greatest effect on the activity of the enzyme?

c) With reference to this experiment only, suggest ONE reason why people are advised against drinking excessive quantities of alcohol.

d) Express the masses of food digested under conditions of 10, 20 and 30% alcohol as a simple whole number ratio.
e) Calculate the percentage decrease in enzyme activity that occurred between 10 and 20% alcohol.

9 Trypsin is an enzyme made in the human pancreas. Powdered milk suspension is a source of protein with a white cloudy appearance. The accompanying diagram shows an experiment set up to investigate the effect of boiling and the effect of alkali on the activity of trypsin.

a) This investigation consists of two experiments conducted at the same time.
(i) Which TWO tubes should be compared at the end of the experiment to draw a conclusion about the effect of boiling on trypsin's activity?
(ii) Which of these two tubes is the control?
(iii) What conclusion can be drawn about the effect of boiling?

b) (i) Which TWO tubes should be compared at the end of the experiment to draw a conclusion about the effect of alkali on trypsin's activity?
(ii) Which of these two tubes is the control?
(iii) What conclusion can be drawn about the effect of alkali on trypsin's activity?
(iv) In what way do the results support your answer to (iii)?
c) Why would a comparison of tubes A and C be invalid?

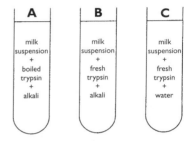

	A	B	C
	milk suspension + boiled trypsin + alkali	milk suspension + fresh trypsin + alkali	milk suspension + fresh trypsin + water
appearance of contents of tube at start	cloudy	cloudy	cloudy
appearance after 1 hour at 37°C	cloudy	clear	slightly cloudy

8 *Investigating aerobic respiration*

1 Match the terms in list X with their descriptions in list Y.

list X

1) aerobic respiration
2) carbon dioxide
3) chemical
4) glucose
5) heat
6) kilojoule
7) lime water
8) oxygen

list Y

a) element essential for efficient release of energy from food during respiration
b) type of energy released during respiration
c) process by which energy is released from food using oxygen
d) chemical solution used to detect the presence of carbon dioxide
e) waste product of respiration
f) fuel-food containing energy
g) term used to describe the type of energy present in food before respiration takes place
h) unit used to measure energy

2 The experiment shown in the diagram below was used to compare the amount of heat energy released by two foodstuffs. State TWO factors that must be kept constant for a valid comparison to be made between the two food samples.

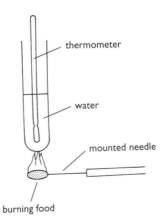

thermometer

water

mounted needle

burning food

Problem Solving in Biology

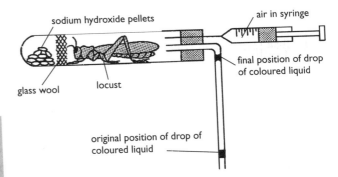

a) Pests X, Y and Z were found at the points shown. In general, what has happened to the temperature of the grain at each of these points? Explain why.

b) Which pest was found nearest to the surface of the grain?

c) Which pest was most widespread in its vertical distribution in the grain?

d) Predict the form that the last part of the line graph would take if the grains at depth 550–700 mm were damp and had begun to germinate. Explain your answer.

sodium hydroxide pellets

air in syringe

final position of drop of coloured liquid

glass wool

locust

original position of drop of coloured liquid

9 A boy set up the experiment above to investigate gas exchange in a locust.

a) What gas, breathed out by the locust, is absorbed by the sodium hydroxide pellets?

b) The boy claimed that the experiment shows that a locust breathes in oxygen. Explain why he was justified in making this claim.

c) Using only the apparatus shown, suggest how the boy could measure the volume of oxygen taken in by the locust during a given period of time.

d) Why is it essential to keep the entire apparatus at constant temperature throughout the experiment?

3 The World of Plants

9 Introducing plants

type of plant		
	herb	
	tree	
	cactus	

abundance of plant		
	•	nearing extinction
	• •	rare
	• • •	common

geographical location		
		northern hemisphere only
		southern hemisphere only
		world-wide distribution

chemicals present in tissues		
		poisons
		drugs
		fuel hydrocarbons

ecological relationships		
		only home of very rare monkey
		flowers pollinated by bats
		nectar eaten by rare butterfly
	?	unknown at present

1 The symbols given in the table are used to describe plants A–L listed below.

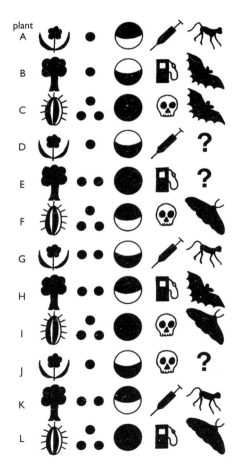

a) One of the above plants is a common, poisonous cactus with a world-wide distribution whose flowers provide food for a rare species of butterfly. Which one?

b) Which herb plant, found only in the southern hemisphere, is nearing extinction yet if saved may prove to be of medical use to mankind in the future?

c) Which rare plant, found only in the northern hemisphere, provides the habitat for an endangered animal species?

d) Identify the plant that contains fuel hydrocarbons yet is nearing extinction.

e) Describe plant J.

f) Name THREE features that are shared by plants A and G.

g) Compare plants C and H in detail.

plant	edible part	% protein content
broad bean	seed	7
almond	nut	20
lentil	seed	24
apricot	fruit (dried)	5
pea	seed	6

3 a) Draw a bar graph of the above information and insert the name of each edible plant part inside its own bar.

b) By how many times is the percentage of protein in almond nut greater than that present in dried apricot fruit?

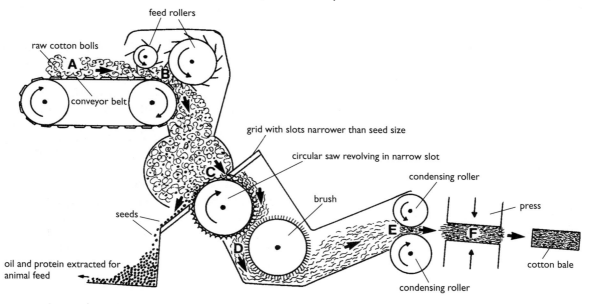

2 Cotton bolls contain tiny seeds entangled in masses of woolly-looking fibres called lint. Separation of seeds from lint is done by a machine called a gin as shown in the diagram above.

a) Which letter in the diagram indicates the point in the process where the cotton fibres are caught on teeth and pulled away from the seeds?

b) To what use are the seeds later put?

c) How many different types of rollers are named in the diagram?

d) Would the brush be turning clockwise or anticlockwise?

e) What finally happens to the brushed cotton fibres in the above process?

4 Read the following passage.

Two parts of a plant provided food for animals: the easily-digested cell contents and the indigestible cellulose cell walls. Most herbivorous (plant-eating) mammals overcome the cell wall 'problem' by having long guts inhabited by cellulose-digesting bacteria.

The panda is a herbivore with a relatively short gut that lacks bacterial assistants capable of breaking down plant cell walls. The panda has to gain most of its food from the cell contents of bamboo shoots and other plants that it eats. To survive it has to work continuous shifts of 8 hours of feeding and 4 hours of sleeping, day and night.

In the cold, damp **Chinese** forests where the panda lives, bamboo plants produce only shoots for long periods of time (i.e. 60 years) and then flower, produce seeds and die. It takes up to 15 years for the new plants to develop shoots that the panda can eat.

In earlier times pandas were able to migrate from one wooded hillside to another where flowering had yet to take place. However, within the last few years their ways have been barred by man clearing and cultivating the floors of the wooded valleys. Even the **Wolong Panda Reserve**, one of the animal's few remaining strongholds, is steadily shrinking under the ever increasing demands of local people for firewood.

a) State ONE way in which the structure of the gut of a panda differs from that of most herbivorous mammals.
b) What benefit is gained by herbivores that have bacterial 'assistants' in their gut?
c) Explain fully why a panda must spend so much of its time feeding.
d) Why do pandas suddenly have to move from one place to another after having lived there for many years?
e) What has prevented such a migration in recent years?
f) If the present situation continues, predict the fate of the panda.

	WEST ⟵		EAST ➝
year	state X	state Y	state Z
1946	10.3%	9.4%	8.9%
1948	10.7	9.7	9.6
1950	11.9	11.5	10.1
1952	12.4	12.2	10.6
1954	12.9	12.6	11.2

5 The figures in the above table show the average percentages of protein present in one species of cereal plant growing in three different states of the USA.
   Draw TWO conclusions from the information.

6 Read the following passage and study the diagrams that relate to it.

In 1987 the yeheb bush (*Cordeauxia edulis*) only grew wild in a few poor regions of **Africa** where many of the people were starving. Many years previously, the plant had been wide-spread in the **Horn of Africa** and its nuts had been a traditional food of the local nomadic people. In addition the plant had been an abundant source of dyes, fuel and forage for live-stock.

Of the few plants that remained in 1987, demand greatly outstripped supply; any nuts that did develop were soon eaten and the vegetation was overgrazed by goats. If this situation had been allowed to continue, the yeheb plant could have faced extinction.

a) Name TWO countries in which wild populations of yeheb plant were still found to be growing in 1987.
b) Describe yeheb's natural habitat.
c) (i) Give FOUR traditional uses of the yeheb bush.
(ii) According to the passage, which TWO of these were threatening the plant's very existence in 1987?

yeheb bush

2–3 metres

dry soil receiving annual rainfall of less than 250 mm

horn

pod

nut

food type present in nut	% content of nut
sugar	12
protein	13
starch	40
fat	12

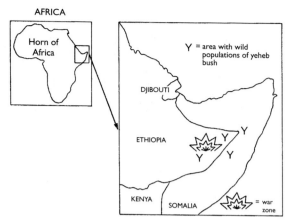

AFRICA

Horn of Africa

Y = area with wild populations of yeheb bush

DJIBOUTI

ETHIOPIA

Y Y Y Y

KENYA

SOMALIA

= war zone

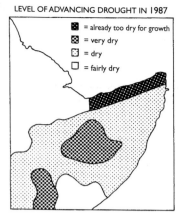

LEVEL OF ADVANCING DROUGHT IN 1987

= already too dry for growth
= very dry
= dry
= fairly dry

(iii) From the maps identify two other problems that may have been causing a reduction in the number of wild yeheb bushes in 1987.

d) Construct a bar graph of the contents of a yeheb nut.

e) If an average nut weighs 1.6 g, calculate the mass of starch that it contains.

f) An Ayrshire potato tuber contains on average 19% starch, 2% protein, 0.5% sugar and no fat. Would the potato be preferable to the yeheb nut as the main part of the Horn of Africa people's diet? Explain your answer.

g) Suggest why the local people did not plant the few remaining yeheb nuts and grow a crop in 1987.

h) The yeheb has now been introduced to Sudan, Yemen and Kenya. If this experiment proves to be a success, what conservation measure should be adopted when peace returns to the Horn of Africa?

7 Read the following passage.

In one particular year the world produced 300 million metric tonnes of both wheat grains and potato tubers, making them appear to be of similar importance as food crops.

However, the average water content of a potato tuber is 80% of its mass whereas that of a wheat grain is only 12%.

In addition, during the preparation of potatoes, often as much as 25% is removed during peeling and thrown away whereas the 25% of wheat grains removed during milling is used in animal feed.

a) Calculate the total weight of water present in the world crop of (i) wheat grains, (ii) potato tubers, during the year referred to in the passage.

b) Give TWO reasons why, weight for weight, wheat is a more valuable crop than potato.

c) A food scientist began with 1000 g of potato tubers. He peeled them all, thus losing 25% of their mass, and then he dried the potatoes to constant weight to form potato powder. Calculate the weight of potato powder formed.

8 The graph and key below refer to the economics and management of a forest of conifers in Scotland.

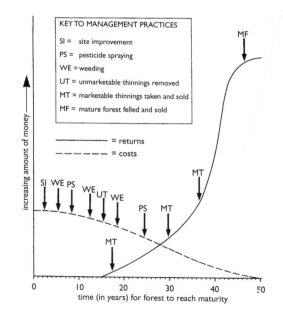

KEY TO MANAGEMENT PRACTICES

SI = site improvement
PS = pesticide spraying
WE = weeding
UT = unmarketable thinnings removed
MT = marketable thinnings taken and sold
MF = mature forest felled and sold

———— = returns
‑ ‑ ‑ ‑ = costs

increasing amount of money

SI WE PS   WE UT WE   PS   MT   MT

MT

MF

0   10   20   30   40   50
time (in years) for forest to reach maturity

a) How many times was the crop sprayed with pesticide?
b) How many different types of management practice were applied to the forest during its first 25 years of growth?
c) In which year did costs equal returns?

d) In general, what relationship exists between costs and returns over the given period of time? Explain why this is the case.
e) Suggest why weeding and thinning out unhealthy weaker trees improves the quality of the remaining trees in the long run.

10 *Growing plants*

I Match the terms in list X with their descriptions in list Y.

list X	**list Y**
I) anther	**a)** structure containing embryo plant and food store formed from an ovule following fertilisation
2) fertilisation	**b)** structures present in a flower's anthers that contain male gametes
3) fruit	**c)** structure present in a flower's ovary that contains female gamete
4) germination	**d)** structure that protects unopened floral bud
5) ovary	**e)** structure containing one or more seeds
6) ovule	**f)** head of stamen containing pollen grains
7) petal	**g)** process by which a male gamete fuses with a female gamete
8) pollen grains	**h)** cell formed when a female gamete is fertilised by a male gamete
9) pollination	**i)** development of a plant embryo into an independent plant with green leaves
10) seed	**j)** brightly coloured scented structure which attracts insects to a flower
11) sepal	**k)** swollen region of flower's female sex organ containing ovules
12) zygote	**l)** transfer of pollen grains from an anther to a stigma

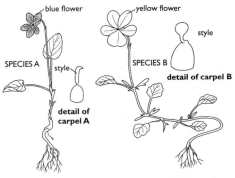

2 a) Use the following key to identify the above two species of *Viola*:

1 style expanded into ball-like stigma 2
style not expanded into ball-like stigma 3

2 long creeping horizontal stem present. *Viola lutea*
no long creeping horizontal stem present ....*Viola tricolor*

3 style extended into hook-like stigma 4
style not extended into hook-like stigma.............
....*Viola palustris*

4 leaf stalks (petioles) hairy 5
leaf stalks (petioles) not hairy...............................6

5 flowers sweetly scented ....*Viola odorata*
flowers not sweetly scented ....*Viola hirta*

6 petals blue...*Viola canina*
petals white ....*Viola stagnina*

b) Describe *Viola odorata*.

3 Read the following passage.

Following pollination, the flowers of *Banksia* (a flowering plant of the Australian bush) die. However, instead of dropping off, the seeds

Problem Solving in Biology

remain firmly attached to the parent plant for years waiting for the arrival of an agent which is disastrous to most living things – a bush fire.

Banksia seeds not only survive fire but actually depend on it to open their seed coats. A seed case consists of two woody valves hinged together and it only opens when it has been thoroughly desiccated by fire.

Once the fire has passed, the paper-thin, single-winged seeds need wind to disperse them among the ashes where they germinate when the rains arrive.

a) Choose the most suitable title for the passage from the following:

A Seed dispersal in Australian flowering plants
B The role of fire in *Banksia's* life cycle
C Flammable agents of seed dispersal in flowers
D *Banksia's* dependence on fire for pollination

b) Choose TWO phrases from the passage that show *Banksia* to be very different from other flowering plants.

4 One hundred different species of flowering plants were investigated to find out which methods could be successfully used to propagate them artificially. The following results were obtained.

number of plant species	method of artificial propagation	code
25	grafting	G
5	layering	L
50	cutting	C
20	tissue culture	T

Which of the following pie-charts correctly represents this information?

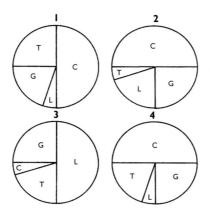

5 Match the terms in list X with their descriptions in list Y.

list X	list Y
1) artificial propagation	**a)** a stump of less desirable but hardy variety of plant to which a scion is grafted
2) asexual reproduction	**b)** general term used to refer to reproduction which involves two parents and the fusion of two sex cells
3) taking cuttings	**c)** natural form of asexual reproduction where the parent plant produces side branches bearing young plants
4) grafting	**d)** general term used to refer to asexual reproduction where the plant is used by humans to produce offspring by artificial means
5) runner formation	**e)** part of a desirable plant used during grafting
6) scion	**f)** natural form of asexual reproduction where parent plant produces storage organs at the ends of underground stems
7) sexual reproduction	**g)** general term for production of identical offspring by one parent without the involvement of sex cells
8) stock	**h)** form of artificial propagation involving removal and planting of side shoots from a parent plant
9) tuber formation	**i)** form of artificial propagation involving the union of stock and a scion

6 Read the following passage.

Although often confused with the coconut palm, the cocoa (cacao) tree is an entirely different plant. When mature it is an evergreen tree with large leaves and pinkish–yellow flowers that develop into pods. Each pod contains about 40 cocoa beans, which are full of useful foodstuffs.

On harvesting, the beans must be fermented, dried, roasted and crushed. Then their contents can be made into highly nourishing cocoa powder (drinking chocolate). Milk chocolate is a mixture of cocoa powder, sugar and milk. The main producers of cocoa are the West Indies and Central and South America.

a) Choose a suitable title for the above passage from the following:
A Types of useful bean plant
B Palm oil from coconuts
C Production of cocoa powder
D Chocolate for a balanced diet
b) Identify TWO phrases in the passage that tell you the cocoa plant is of importance in human nutrition.
c) The average annual yield of a cocoa tree is found to be 25 pods. If the total numbers of beans from 25 healthy pods weigh 1000 g, what is the average mass of one fresh cocoa bean?

d) If only 0.4 g of a cocoa bean's mass is converted into cocoa powder, calculate in kilograms the average yield of cocoa powder for a plantation of 200 trees where a quarter of the tress are too young to produce pods.
e) In a diseased plantation, the total number of beans from 25 pods only weighed 505 g. Express this as a percentage of the weight of beans from 25 healthy pods.

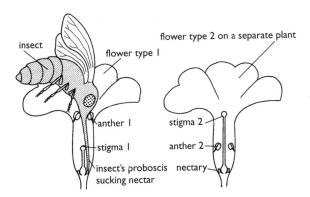

7 Cross-pollination is the transfer of pollen from the anther of one flower to the stigma of another flower on a separate plant of the same species.
a) Explain the difference between cross- and self-pollination.

b) with reference to all of the named structures in the accompanying diagram, describe how cross-pollination takes place between flower types 1 and 2.

8 The following table gives the optimum, maximum and minimum temperatures at which the grains or seeds of three types of plant germinate.

plant	optimum (°C)	minimum (°C)	maximum (°C)
maize	37–44	8	44
wheat	25–30	3	35
cucumber	31–37	15	49

a) How many of the plants could be germinated at 12°C?
b) Which plants could be germinated at 47°C?

c) Could all three plant types be germinated within their optimum range of temperature simultaneously in one greenhouse? Explain your answer.

9 The diagram below and overleaf shows six methods of asexual reproduction in plants.
a) From the information in the diagram state TWO differences between a bulb and a corm.
b) Identify ONE feature shared by both a tuber and a rhizome.

c) Construct a key of paired statements that allows each method of asexual reproduction to be correctly classified. (Use only the information given.)

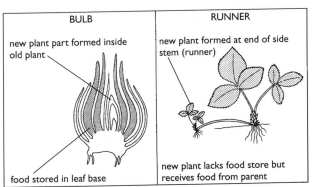

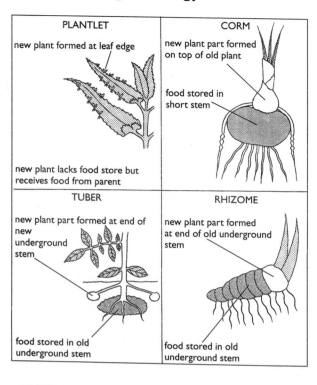

PLANTLET	CORM
new plant formed at leaf edge	new plant part formed on top of old plant
	food stored in short stem
new plant lacks food store but receives food from parent	
TUBER	RHIZOME
new plant part formed at end of new underground stem	new plant part formed at end of old underground stem
food stored in old underground stem	food stored in old underground stem

10 One method of natural vegetative propagation amongst plants is bulb formation. As part of an investigation of bulbs, a girl was given an onion bulb and asked to test its fleshy storage leaves for simple sugar and starch. She found sugar to be present and starch to be absent.

She concluded that all bulbs store sugar but not starch in their leaf bases.

a) Why was she not justified in drawing this conclusion?

b) What information would the girl need to have before attempting to draw a conclusion about the type of food stored in bulbs?

---

11 *Making food*

1 Match the terms in list X with their descriptions in list Y.

list X

1) carbon dioxide
2) chlorophyll
3) glucose
4) guard cell
5) oxygen
6) phloem
7) photosynthesis
8) starch
9) stomata
10) xylem

list Y

a) tissue that transports water in a plant
b) gaseous by-product of photosynthesis released by plant
c) tiny pores found on the surface of a leaf through which gases pass
d) process by which a green plant converts light energy to chemical energy stored in food
e) complex carbohydrate formed from many glucose units linked together
f) one of a pair that control gas exchange through a stoma
g) tissue that transports soluble food in a plant
h) simple sugar formed during photosynthesis in green plants
i) gas essential for photosynthesis, which enters leaves via stomata
j) green pigment that traps light energy during photosynthesis

2

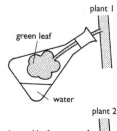

plant 1

green leaf

water

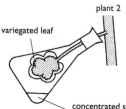

plant 2

variegated leaf

concentrated sodium hydroxide solution (absorbs CO_2)

The experiment opposite is an alternative method of demonstrating that carbon dioxide is needed by a green plant for photosynthesis.

State TWO ways in which this experimental set-up would have to be altered to allow a valid comparison to be made.

3 A variegated ivy plant was given all of the conditions needed for photosynthesis. When one of its leaves was tested for starch, the result shown in the table below was obtained. It was therefore concluded that chlorophyll is required for photosynthesis.

a) Copy and complete the table to show the appearance of the other two leaf types under the same conditions after the starch test.

b) Imagine that a cutting from the wandering sailor plant had been kept in darkness for two days. Draw a labelled diagram of how one of its leaves would appear after the starch test.

type of variegated plant	before starch test	after starch test
ivy	non–green, green	blue–black, non–blue–black
wandering sailor		
geranium		

4 A scientist grew specimens of a species of cereal plant at different light intensities for several hours and then measured the mass of sugar produced at each light intensity as shown in the following table.

light intensity (foot candles)	sugar production (g/kg of dry leaves)
0	0
100	1.7
200	3.3
300	4.8
400	5.7
500	6.0
600	6.0
700	6.0

a) Using graph paper, construct a line graph of the above results. (Use the horizontal axis for light intensity and the vertical axis for sugar production.)

b) Express the mass of sugar produced per kilogram of dry leaves at light intensity 700 foot candles as a percentage of the dry mass of leaves.

GRASS PLANTS

time on 24 hour clock	CO_2 conc. in air between leaves
10.00	high
13.00	low
16.00	medium

WATER PLANTS

time on 24 hour clock	CO_2 conc. in water between leaves
10.00	high
13.00	medium
16.00	low

5 It was concluded from the results in the above experiment that grass plants were absorbing more CO_2 at 13.00 hours that at either of the other two times.

a) At which of the three times was photosynthesis occurring at the greatest rate in water plants?

b) The pH value of a liquid is known to decrease as its CO_2 concentration increases. At which of the three times will the pond water be at the lowest pH?

Problem Solving in Biology

6 In an experiment to investigate water loss in two
types of plant, leaves were attached to
simple balances as shown in the following diagrams.
a) Which leaf lost most water in experiment 1?
b) Which leaf lost more water in experiment 2?
c) What can be concluded about the location of
most or all of the stomata in a waterlily leaf from
the above experiment?

	at start	after several hours
EXPT.1	privet leaf with lower surface coated with Vaseline privet leaf with upper surface coated with Vaseline pivot	
EXPT.2	waterlily leaf with lower surface coated with Vaseline waterlily leaf with upper surface coated with Vaseline pivot	

7 The following graph shows the effect of increasing
light intensity on the uptake of carbon dioxide by a
green plant.

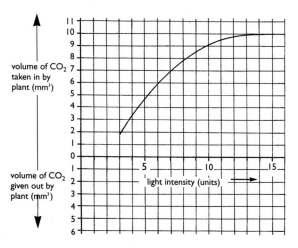

a) How many units of light was the plant receiving
when it took in 6 mm³ of CO_2?
b) What volume of CO_2 was taken in by the plant
at a light intensity of 7 units?
c) What volume of CO_2 was taken in by the plant
at a light intensity of 14 units?

+

d) With the aid of a ruler work out the light
intensity at which CO_2 is neither taken in nor
given out by the plant.

8 The table below gives the results from an investigation into the possible relationship between monthly rainfall and photosynthesis (measured as dry mass of a standard number of leafs discs).

month	rainfall (cm)	dry mass of leaf discs (g)
January	20	2.5
February	14	3.0
March	12	3.5
April	10	4.0
May	8	4.0
June	5	4.5
July	3	5.0
August	4	5.0
September	8	4.5
October	14	4.0
November	19	3.0
December	20	2.0

a) Name the THREE driest months of the year.
b) How many months were wetter than April?
c) During which months did most photosynthesis occur?

d) Present the data graphically by first drawing a bar graph of rainfall and then a line graph of dry mass of leaf discs using a common horizontal axis to highlight the possible relationship.

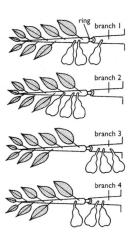

9 The diagrams show four branches of a pear tree at the start of a ringing experiment where the fruit grower is attempting to produce unusually large pears.
   The table below shows the total mass of the three pears per branch several months later. Copy the table and complete it by matching the four sets of results with the four branches.

total mass of three pears per branch	branch number
75g	
300g	
525g	
750g	

Animal Survival

12 *The need for food*

I Match the terms in list X with their description in list Y.

list X

1) alimentary canal
2) amylase
3) canine
4) carnivore
5) digestion
6) enamel ridges
7) herbivore
8) large intestine
9) lipase
10) oesophagus
11) omnivore
12) pepsin
13) small intestine

list Y

a) breakdown of large particles of food into smaller particles

b) animal whose dentition is suited to a mixed diet of plant and animal material

c) enzyme that promotes the digestion of protein

d) animal whose dentition is suited to a diet of plant material only

e) region of the gut from which the end products of digestion are absorbed into the bloodstream

f) muscular tube running through the human body from mouth to anus

g) enzyme that promotes the digestion of fat

h) structures present on a sheep's molar teeth that enable it to grind grass efficiently

i) large teeth in a dog's dentition that enable it to stab and grip prey

j) enzyme that promotes the digestion of starch

k) region of the gut where water is absorbed from waste material

i) animal whose dentition is suited to a diet of animal material only

m) muscular part of alimentary canal connecting the mouth to the stomach

2

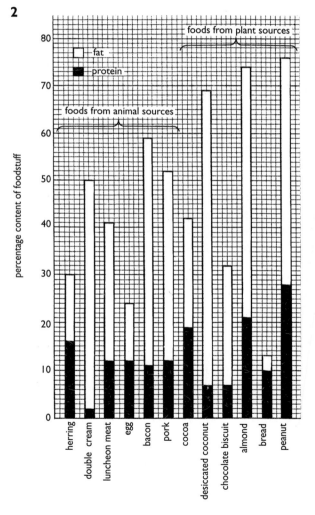

Using only the information given in the bar graph opposite, answer the following questions.
a) Which THREE animal foods contain the same percentage of protein?
b) Which plant foods contain a higher percentage of protein than bread?
c) Name the animal foods that contain a lower percentage of fat than pork.
d) How many foods contain more fat than protein?
e) Which of the following contains least protein? herring, cocoa, almond, peanut.
f) Name THREE foods that contain an equal percentage of fat.

3 a) Calculate the average fat content for the foods shown in the table below.

food	fat content (g/kg)
butter	810
cheese	345
herring	185
lamb	302
milk	38

b) Calculate the average protein content for the foods shown in table below.

food	protein content (g/kg)
beans (baked)	5.1
beef	18.2
cheese	25.5
milk	3.3
peanuts (roasted)	28.2
peas	5.8
potatoes	2.1

4 **A human's first set of teeth are called milk or deciduous teeth. The first milk teeth to appear through the gums are the incisors at about 6–10 months after birth. The full set is normally present by about 2 years but always lacks molar teeth. The first permanent teeth (of the second set) appear at the age of about 6 years. These are the front molars. The milk teeth are then replaced by further permanent teeth. Incisors grow in followed by the premolars and then the canines. Later at about 12 years, the second molars appear and finally the third molars ('wisdom' teeth) at around 18 years or even later giving a full set of 32 teeth.**

a) What is the main theme of the above passage?
b) Identify TWO differences between first and second sets of teeth.
c) List the order in which the permanent teeth appear.
d) How many teeth are present in a full set of milk teeth?

5 The following diagram charts the progress of various foodstuffs along the human gut (alimentary canal).

type of food at start	in mouth	in stomach	in small intestine	substance(s) present after digestion
fat				fatty acids + glycerol
starch				sugar
vitamins				vitamins
protein				amino acids

▒ = undigested food   □ = enzyme-digested food

a) Which type of food remains undigested as it passes through the alimentary canal?
b) Which type of food undergoes digestion in both mouth and small intestine?
c) Name the type of food that is digested in the small intestine only.
d) How many different types of food undergo digestion in the stomach?

6 The chart in the diagram overleaf can be used to try to relate certain symptoms and ailments to possible deficiencies in the human diet.
a) Over 200 years ago, sailors on extended sea voyages often used to suffer soft, bleeding gums. Although the texture of their skin remained normal, any cuts that they received failed to heal. Use the chart to identify the vitamin in which they were deficient and suggest a remedy.
b) If a group of people develop rough, dry skin and have great difficulty seeing in very dim light, which vitamin might their diet be deficient in?

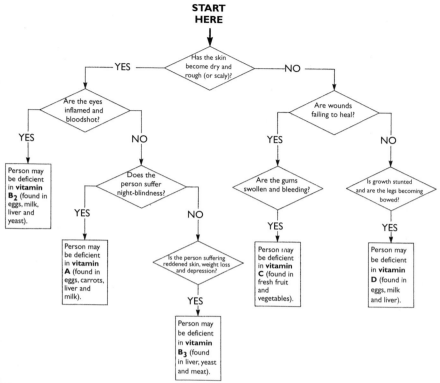

START HERE

Has the skin become dry and rough (or scaly)?
— YES / NO —

YES branch:
Are the eyes inflamed and bloodshot?
- YES → Person may be deficient in **vitamin B₂** (found in eggs, milk, liver and yeast).
- NO → Does the person suffer night-blindness?
  - YES → Person may be deficient in **vitamin A** (found in eggs, carrots, liver and milk).
  - NO → Is the person suffering reddened skin, weight loss and depression?
    - YES → Person may be deficient in **vitamin B₃** (found in liver, yeast and meat).

NO branch:
Are wounds failing to heal?
- YES → Are the gums swollen and bleeding?
  - YES → Person may be deficient in **vitamin C** (found in fresh fruit and vegetables).
- NO → Is growth stunted and are the legs becoming bowed?
  - YES → Person may be deficient in **vitamin D** (found in eggs, milk and liver).

c) Give ONE difference between the symptoms that indicate a deficiency in vitamin C and those that indicate a deficiency in vitamin D.

d) Rickets is the name given to the deficiency disease suffered by children whose bones are so soft that they bend under the weight of the body.
i) In which vitamin are such children deficient?
ii) Which food should be included in their diet to supply the vitamin?

e) According to the chart which TWO foodstuffs together would provide a rich supply of all five vitamins referred to in the chart?

f) State ONE symptom that people deficient in vitamin A share with those deficient in vitamin B₃.

7 A small sample of the natural liquid present in a pig's gut was taken from each of the four lettered regions indicated in the diagram.

Each sample was put into a small hole in a dish of starch agar. After 24 hours at 37°C, the dish was flooded with iodine solution. The result is shown in the diagram overleaf.
a) What colour results when iodine solution reacts with starch?

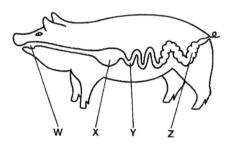

W   X   Y   Z

b) What conclusion can be drawn from the experiment about a pig's ability to digest starch in regions W, X, Y and Z of its gut?
c) Explain why starch digestion is possible in some of these regions of the gut but not in others.

8 Barium sulphate is a chemical that does not allow X-rays to pass through it. The progress of a meal containing barium sulphate can therefore be followed through the alimentary canal using X-rays to detect its location.

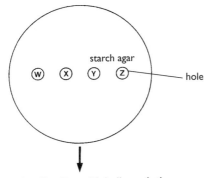

starch agar
w x y z
hole

result after flooding with iodine solution

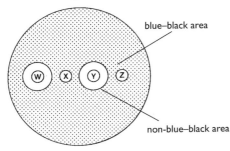

blue–black area
w x y z
non-blue–black area

The first table below shows the arrival and departure times of a barium meal for certain regions of the alimentary canal of six hospital patients (A–F). This information was then used to calculate the time spent by the food in each region. Some of the data is given in the second table below.

	time on 24-hour clock			
patient	arrival of meal in stomach	departure of meal from stomach	departure of meal from small intestine	arrival of meal in rectum
A	06.00	08.19	11.40	23.42
B	06.10	08.05	10.22	22.57
C	06.20	08.20	11.20	22.20
D	06.30	08.27	11.22	22.38
E	06.40	08.52	11.32	21.46
F	06.50	08.57	11.50	23.55

	time (in minutes) spent by meal:		
patient	in stomach	in small intestine	in large intestine
A	139	(i)	722
B	115	137	(ii)
C	120	180	660
D	117	175	676
E	132	(iii)	(iv)
F	127	173	725
average		171	(v)

a) Calculate the average time spent by a barium meal in the stomach of the patients in this survey.
b) Convert patient C's data in the second table from minutes to hours and present the information for the three alimentary canal regions as a pie chart.

c) Calculate the times in minutes that have been omitted from boxes (i) – (v) in the second table.

9 Match the terms in list X with their descriptions in list Y.

list X	list Y
1) amino acids	**a)** energy-rich food such as sugar, starch or glycogen
2) bile	**b)** food composed of amino acids and essential for tissue repair
3) carbohydrate	**c)** chemical element present in protein but absent from carbohydrate and fat
4) fat	**d)** muscular chamber enclosed by two sphincters, which prevent escape of food during churning
5) fatty acid	**e)** end product of carbohydrate digestion
6) glucose	**f)** end products of protein digestion
7) hepatic portal vein	**g)** one of many finger-like projections that increase the absorbing surface of the small intestine
8) lacteal	**h)** type of protease enzyme present in intestinal juice
9) nitrogen	**i)** substance made in liver that converts large drops of fat to small droplets
10) peptidase	**j)** type of protease enzyme present in pancreatic juice
11) peristalsis	**k)** tiny lymphatic vessel that picks up and transports digested fat
12) protein	**l)** energy-rich food composed of fatty acids and glycerol
13) stomach	**m)** an end product of fat digestion
14) trypsin	**n)** blood vessel that transports glucose and amino acids to the liver
15) villus	**o)** wavelike motion of gut wall caused by alternate contraction and relaxation of muscle

10 Choose suitable forms of calculation and then copy and complete the final column in the following table. +

vitamin	minimum daily requirement (mg)	notes	least amount of food (g) that will supply minimum daily requirement of vitamin
A	0.75	166 g margarine contains 1.5 mg of vitamin A	
B_1	1.2	525 g wheat germ contains 3.6 mg of vitamin B_1	
B_{12}	0.003	800 g cheese contains 0.012 mg of vitamin B_{12}	
C	30.0	100 g lettuce contains 15 mg of vitamin C	
D	0.0025	50 g sardines contains 0.0025 mg of vitamin D	

11 The following table gives a summary of a series of experiments in which tough meat was treated with the juice squeezed out of the pawpaw fruit (which grows on palm-like trees in tropical countries).

	age of fruit		state of fruit		pH conditions		resultant state of meat
	mature	immature	boiled	unboiled	neutral	acidic	
experiment 1	✓		✓		✓		tough
2		✓	✓		✓		tough
3	✓			✓	✓		tough
4		✓		✓	✓		tender
5	✓		✓			✓	tough
6		✓	✓			✓	tough
7	✓			✓		✓	tough
8		✓		✓		✓	tough

Study the information carefully then construct a hypothesis to account for the results.

13 *Reproduction*

1 Match the terms in list X with their descriptions in list Y.

list X
1) amnion
2) courting
3) egg
4) embryo
5) fertilisation

list Y
a) male sex cell
b) part of male reproductive system that produces sperm
c) part of female reproductive system that produces eggs
d) fertilised egg
e) mating behaviour that brings male and female animals together and increases chance of fertilisation

6) foetus
7) gestation
8) ovary
9) oviduct

f) tube containing blood vessels that connects mammalian embryo to placenta
g) liquid-filled sac that bathes and cushions mammalian embryo
h) period of development of a mammalian embryo inside the mother's body
i) process by which a sperm's nucleus fuses with that of an egg forming a zygote

10) placenta   **j)** multicellular structure formed from a zygote by repeated cell division

11) sperm   **k)** embryo whose species can be determined from its appearance

12) testis   **l)** disc-shaped organ that allows selective exchange of materials between bloodstream of the embryo and mother

13) umbilical cord   **m)** region of female reproductive system in which fertilisation takes place

14) uterus wall   **n)** female sex cell

15) zygote   **o)** region of female mammal's reproductive system to which the growing embryo becomes attached

2   The following table lists the length of gestation (pregnancy) for several mammals.

mammal	average length of gestation period (days)
elephant	665
rhinoceros	551
horse	338
human	266
chipanzee	227
fox	52
mouse	19

a) By how many days is the gestation period of a horse longer than that of a chimpanzee?
b) Which mammal's length of pregnancy is 29 times longer than that of a mouse?

c) By how many times is the gestation period of an elephant longer than that of a human?
d) Which animal's length of pregnancy is shorter than that of a horse by 6.5 times?
e) What apparent relationship exists between body size and length of gestation period? Suggest why.

3   Imagine that you are the parent of a 3-month-old baby girl who receives a feed every 2–3 hours. The baby has a runny nose and has been coughing but not crying much. Her temperature is 37°C and her faeces are normal but she has just brought up her entire feed for the first time. When you look up

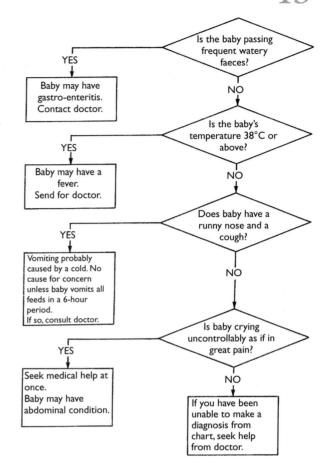

the family medical book for help you find the above chart.
a) What seems to be wrong with the baby?
b) Imagine that you have decided to take the advice given in the flow chart. Describe what you would do over the next few hours.
c) If you read a flow chart like this one but fail to understand it or fail to find a satisfactory answer, what should you do?

4   Read the following passage.

A few hours after birth, a human baby starts to suckle at the mother's breast or from a bottle. The mother's milk contains everything that the baby needs except iron. A store of this chemical element has already been obtained from the mother's bloodstream by the foetus during gestation. This supply lasts for a few months after birth.

Problem Solving in Biology

Bottle-fed babies receive dried cow's milk that has been altered in advance by the manufacturers since natural cow's milk contains more protein and less vitamin A and sugar than human milk. If strict hygiene is not observed, bottle-feeding can pass germs on to the baby. Human milk is free of germs and in addition contains antibodies, which defend the baby against disease. Dried cow's milk lacks these human antibodies.

The crying response in babies is present from birth. It consists of muscular activity, exaggerated breathing, watering of the eyes and high pitched sounds. Almost all young mammals cry when insecure, frightened or in pain. By crying the young animal signals to its parents that something may be wrong. If it continues to cry, the parent will normally examine the young animal for sources of pain or injury.

a) Give TWO changes that manufactures have to make to cow's milk to make it suitable for human consumption.

b) From where does a human baby obtain iron during the first few months after birth?

c) In addition to containing exactly the right balance of foodstuffs, state TWO other ways in which human milk is better than cow's milk for infants.

d) State the THREE situations given in the passage that can make a baby cry.

e) Name the main signal present in the crying response that quickly draws the parent's attention to the infant.

5 The following diagram gives six pieces of information about each of nine different animals native to North Island, New Zealand. Draw up a table to display this information clearly.

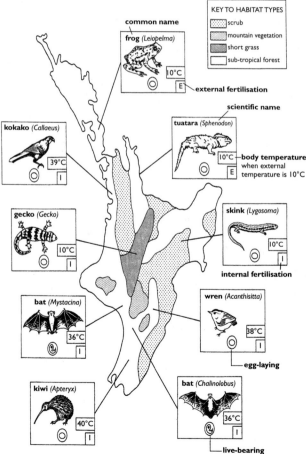

6 The following graph represents the results from a survey done to investigate the number of twin births that occur per 1000 pregnancies for women of different ages in Britain.

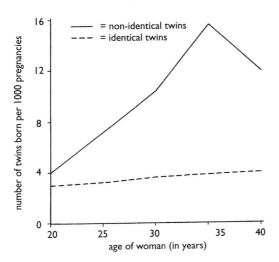

a) Suggest why as many as 1000 pregnancies were studied in the survey.
b) Women from all walks of life (i.e. socio-economic groups) were included. Why?
c) By comparing the line graphs, draw TWO appropriate conclusions from the results.

7 The following table refers to the survival of several types of young vertebrate animals.

a) Calculate the values omitted from boxes (i), (ii) and (iii) in the table.
b) (i) What relationship exists between the number of eggs produced annually and the length of parental care given to the young?
(ii) What relationship exists between percentage survival of young and length of parental care?
c) Suggest why some of the information in the table can only be given as approximate values.

8 One hundred eggs are present in each clutch laid by a green turtle. Assume that she lays four clutches at 9-day intervals on the same beach. Monitor lizards then dig up and eat 40% of the eggs. All the remaining eggs hatch out but 80% of the young turtles are caught by gulls and fiddler crabs as they make their way across the beach. Calculate how many baby turtles reach the sea.

9 A shepherd recorded the average mass of the lambs in his flock in the spring of 1996. Some ewes only gave birth to a single lamb, which they reared themselves. Some gave birth to twins. Of these ewes, some were left to rear both of their twins normally; others were allowed to rear a 'single' twin and a ewe with no lambs was given the other 'single' twin to rear.

Here are some extracts from the shepherd's diary.

animal	class of vertebrate	number of eggs produced annually	number surviving after 1 year	% survival of young	length of time spent by parents feeding and protecting young (weeks)
cod	fish	4 000 000	(i)	0.01	0
trout	fish	3000	150	(ii)	0
sparrow	bird	15	12	80	2
ptarmigan	bird	(iii)	9	90	3
fox	mammal	4	4	100	10
whale	mammal	1	1	100	80

1st March
Many lambs born to-day.
Average mass of single lamb 6 kg.
Average mass of twin lamb 4·5 kg.

29th March
Lambs all doing well despite snow.
Average mass of single lamb 15 kg,
normal twin 10·5 kg and 'single' twin 12·5 kg.

26th April
Snow away and all lambs surviving.
Average mass of single lamb 24 kg,
normal twin 17·5 kg and 'single' twin 21 kg.

24th May
Weather much warmer. Clear differences
in size amongst lambs.
Single lamb now 33 kg, normal twin 24·5 kg
and 'single' twin 29 kg on average.

a) Present this information in a table so that the average mass for each type of lamb as it changes with time can be followed at a glance.
b) Draw TWO conclusions from the information in your table.

14 *Water and waste*

I Match the terms in list X with their descriptions in list Y.

list X	list Y
1) bladder	**a)** solution of urea and salts excreted by the kidneys
2) filtration	**b)** tube that carries urine from kidney to bladder
3) kidney	**c)** blood vessel that brings unpurified blood to a kidney
4) reabsorption	**d)** harmful waste product made in liver
5) renal artery	**e)** organ for temporary storage of urine
6) renal vein	**f)** process by which small molecules pass out of blood into kidney units
7) urea	**g)** process by which useful molecules are returned from kidney tubules to the bloodstream
8) ureter	**h)** tube that allows the exit of urine from the bladder to the external environment
9) urethra	**i)** one of two bean-shaped organs that excrete urine and maintain water balance
10) urine	**j)** blood vessel that takes purified blood away from kidney

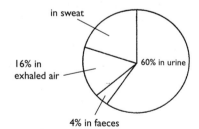

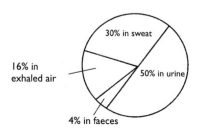

50

2 The first pie chart shown represents a person's daily water loss.
a) What percentage of this person's water was lost as sweat?
b) If the total volume of water lost was 2500 cm³, what volume of water was lost in exhaled air?
The second pie chart represents the same person's daily water loss at a different time of year.
c) Suggest what environmental factor was responsible for the difference between the two pie charts. Explain your answer.

3 The results given in the following table are from an experiment to investigate the response of a man to drinking 1 litre of water. Urine was collected at half-hourly intervals.

time (min)	urine output (cm³/30 min)
start (drink given)	50
30	340
60	490
90	160
120	40

a) Using the axes given opposite and similar graph paper, present the above data as a bar graph.
b) Calculate the total volume of urine collected after the experiment had been running for 90 minutes.

4 The table below shows some of the results from an experiment set up to calculate the water content of three common foods.

	cornflakes	lettuce	bread
mass of empty dish (g)	15	15	15
mass of dish + sample of fresh food (g)	35	40	39
mass of fresh food (g)	20	(ii)	24
mass of dish + sample of dry food (g)	34	16	30.6
mass of dry food (g)	19	1	(iv)
mass of water lost from food (g)	1	24	(v)
% water present in fresh food	(i)	(iii)	(vi)

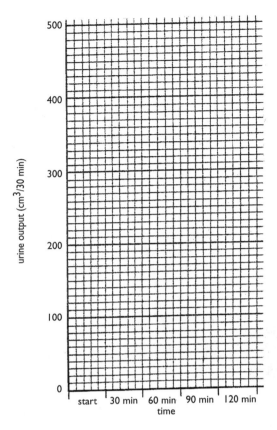

urine output (cm³/30 min)

start   30 min   60 min   90 min   120 min
time

a) Calculate the answer missing from boxes (i), (ii) and (iii).

b) Calculate the answers missing from boxes (iv), (v) and (vi).

5 The following tables show the water balance figures for a human subject who remained under constant environmental conditions for two days.

water gain	day 1	day 2
in food	850 cm³	850 cm³
in drink	1300 cm³	1600 cm³
formed during respiration	350 cm³	350 cm³
total gain	2500 cm³	2800 cm³

Problem Solving in Biology

water loss	day 1	day 2
from lungs		
from kidneys	1500 cm³	
from skin	500 cm³	500 cm³
in faeces	100 cm³	100 cm³
total loss	2500 cm³	2800 cm³

a) In what way did water gain differ between day 1 and day 2?

b) Calculate how much water was lost in exhaled air during day 1.

c) Calculate the volume of urine expelled during day 2.

d) From which TWO parts of the body would loss of water have increased if the human subject had exercised vigorously during part of the investigation?

6 The enzyme urease catalyses the following reaction:

urea + water $\xrightarrow{\text{urease}}$ ammonia + carbon dioxide

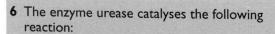

Ammonia is a gas which makes moist red litmus paper turn blue.

In an experiment two test tubes were set up as shown in the first of the accompanying diagrams and placed in a water bath at 35°C. After 27 minutes the litmus paper above X changed colour whereas that above Y remained unaltered. The second diagram shows a kidney and associated tubes.

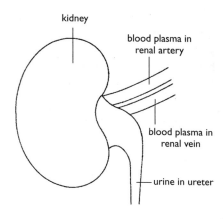

kidney — blood plasma in renal artery — blood plasma in renal vein — urine in ureter

a) (i) Match blood plasma types X and Y with the corresponding blood vessels in the second diagram.

(ii) Explain how you arrived at your answer to (i).

b) Rewrite the following sentences choosing the correct answer from each choice of alternatives. If the experiment was repeated using a third test tube containing a sample of liquid from the ureter of a thirsty mammal, the time taken for the litmus to turn from blue/red to red/blue would be longer/shorter than that taken by the liquid in tube X. This is because an equal volume of urine would contain a higher/lower concentration of urea than liquid X.

c) Predict the effect of repeating the experiment shown in the first diagram in a fridge at 5°C.

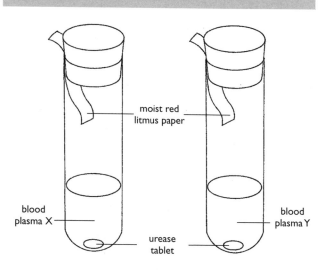

moist red litmus paper

blood plasma X

blood plasma Y

urease tablet

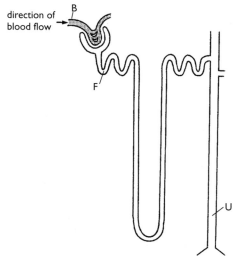

direction of blood flow

B

F

U

7 The diagram at the bottom of page 52 shows a kidney filtering unit (nephron) and part of its blood supply. The concentrations of some of the substances found in the blood (B), the filtrate (F) and the urine (U) are given in the following table.

	urea (g/100 cm³)	glucose (g/100 cm³)	protein (g/100 cm³)	salts (g/100 cm³)
blood (B)	0.03	0.10	8.00	0.72
filtrate (F)	0.03	0.10	0.00	0.72
urine (U)	2.00	0.00	0.00	1.50

a) Which substances did not pass from B to F?
b) Which substance present in the filtrate was completely reabsorbed back into the bloodstream?
c) Which substance became concentrated in urine to the greatest extent?
d) If a man produced 180 litres of filtrate and 1.8 litres of urine in one day, what percentage of filtrate was passed as urine?

8 Read the passage and answer the questions based on it.

In addition to *diabetes mellitus* (sugar diabetes), there exists another disease called *diabetes insipidus*. This is characterised by excessive thirst and the production of large quantities of very dilute urine. The disease can be a malfunction of the posterior lobe of the pituitary gland. This results in insufficient ADH (anti-diuretic hormone) being secreted into the bloodstream and therefore less water being reabsorbed from the kidney tubules back into the blood. Instead vast quantities of dilute urine are produced.

In some cases *diabetes insipidus* is treated by taking ADH in the form of nasal snuff. Without treatment a sufferer has to drink many litres of water in a single day to make up for water losses.

a) Explain why a malfunction of the posterior lobe of the pituitary gland can cause *diabetes insipidus*.
b) Apart from thirst, what is the main symptom of *diabetes insipidus*?
c) Why does lack of sufficient ADH lead to the symptom you gave as your answer to **(b)**?
d) Suggest what could happen if a sufferer of *diabetes insipidus* took an exceptionally high dose of nasal snuff containing ADH.

15 Responding to the environment

I Match the terms in list X with their descriptions in list Y.

list X
1) annual
2) biological clock
3) choice chamber
4) circadian
5) external trigger
6) hibernation
7) migration
8) response
9) rhythmical

list Y
a) type of rhythmical behaviour where an animal survives winter in a state of prolonged inactivity
b) general term for a factor able to bring about a response by an organism
c) reaction shown by an organism on being exposed to a stimulus
d) type of rhythmical behaviour that occurs in response to movement of the sea
e) general term used to describe a behaviour pattern that is repeated at definite intervals
f) type of rhythmical behaviour that occurs once every 24 hours
g) general term referring to rhythmical behaviour that occurs once a year
h) stimulus that initiates rhythmical behaviour in an organism
i) piece of apparatus used to study the effect of differences in an environmental factor on an animal's behaviour

10) stimulus

11) tidal

j) form of annual rhythmical behaviour where the animal moves to a warmer climate in autumn

k) part of an animal's nervous system thought to exert internal control over rhythmical behaviour

e) Why must all the woodlice used be of the same species?

f) In a further experiment the boundary between the damp and dry sides of the choice chamber was blocked with cardboard after 1 minute. The seven woodlice trapped in the dry side eventually slowed down after several minutes and bunched up together against the side of the chamber in one group. Suggest why this response is of survival value to the animals.

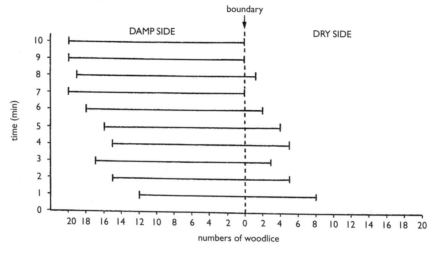

2 Twenty woodlice were released into a choice chamber. The numbers of animals present in both sides were recorded every minute for 10 minutes. The results are shown in the following graph.
a) What was the one variable factor investigated in this experiment?
b) How many woodlice were present on the (i) dry side at minute 5? (ii) damp side at minute 8?
c) At which minute were there (i) 17 woodlice on the damp side? (ii) 2 woodlice on the dry side?
d) What conclusion can be drawn from this experiment about the response of woodlice to the environmental factor under investigation?

3 The following graphs record the activity of three processes that occur in a human body over a period of 3 days.
a) For how many hours did the person sleep each night?
b) With reference to cycle of activity, what generalisation can be made about the three body processes represented in the graphs?

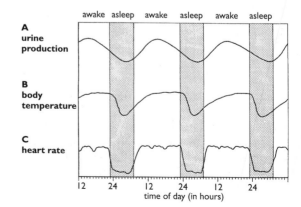

c) Suggest why the rhythmical behaviour pattern shown in graph C is of benefit.
d) Cell division in the skin shows the reverse rhythm to that of heart rate. Between which hours (approximately) would you expect wound healing to be most rapid?

4 The choice chamber shown in the accompanying diagram was set up to study the response of the fruit fly, *Drosophila melanogaster*, to different levels of light intensity.

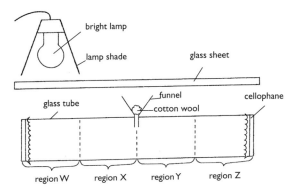

Five trials were run where the lamp was the only light source. Each group of 20 flies was introduced via the funnel and the results noted for each trial after 3 minutes. The results are summarised in the following table.

		number of flies in the region of the tube after 3 minutes					
		trial					average
		1	2	3	4	5	
region	W	18	17	18	18	19	
	X	2	2	1	2	0	1.4
	Y	0	0	1	0	1	0.4
	Z	0	1	0	0	0	0.2

a) Calculate the average value omitted from the results table.
b) Draw a valid conclusion from the results.
c) (i) Identify the means by which a named abiotic factor was prevented from affecting the experiment.

(ii) Why would the experiment have been invalid if this precaution had not been taken?
d) Explain why several trials were run.

e) (i) Express the average number of flies in region Y as a percentage of the total.
(ii) Express, as a whole number ratio, the average number of flies in region W to the total average number in the other three regions.
f) Predict the effect of repeating the experiment with the lamp switched off.

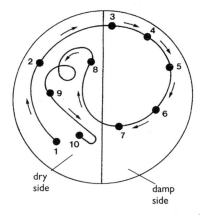

5 The above diagram shows a choice chamber into which a boy dropped a woodlouse through a hole in the lid. The woodlouse landed at position 1. (All holes in the choice chamber's lid were kept closed during the experiment.) The distance between two numbered dots represents the distance covered by the woodlouse every 5 seconds.
a) What was the one variable factor controlled by the experimenter in this investigation?
b) How many conditions of this factor were included?
c) Describe the effect of each of these conditions on the woodlouse's rate of activity.

d) Explain how you arrived at your answer to part c).
e) How long did it take the woodlouse to move from position 1 to position 10?
f) From results, the boy concluded that all woodlice behave in this way. Why was he NOT justified in making such a generalisation?

6 The following experiment was set up to investigate the effect of two environmental factors on the distribution of a tiny animal that feeds on green algae at the surface of the pond where it lives.

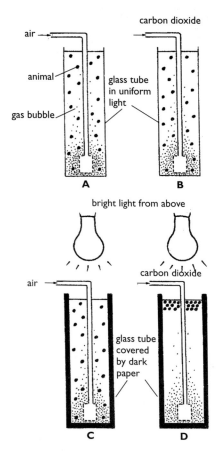

a) Name the one environmental factor by which tubes A and B differ.

b) What effect does this difference have on the distribution of the animals?

c) Name the one environmental factor by which tubes A and C differ.

d) What effect does this difference have on the distribution of the animals.

e) Consider the distribution of the animals in tube D. What conclusion does this result allow us to draw from the experiment?

f) Suggest why the response shown by this animal to its environment is of survival value.

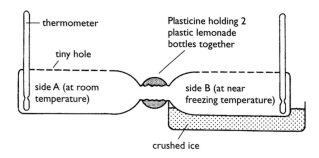

7 The apparatus above was designed by a group of pupils in order to investigate the response of slugs to a choice between room temperature and near freezing conditions. It was their intention to insert 20 slugs of the same species and then use the apparatus exactly as shown over a period of 1 hour without making any alterations.

a) Suggest why they made tiny holes in the lemonade bottles.

b) Consider carefully the aim of the experiment and then state a source of error that will affect the intended attempt described above.

c) Suggest how this source of error could be minimised.

8 Read the following passage.

Migration in birds normally involves an annual round trip between two areas, each of which offers conditions more favourable than the other for part of the year. Some migratory birds move all the way from one hemisphere of the world to the other and back again; others cover less distance by simply migrating to a tropical region within their own half of the world. The British yellow wagtail belongs to this second category as shown by the following map.

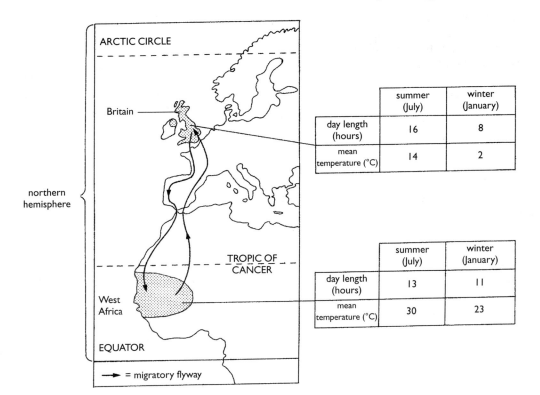

	summer (July)	winter (January)
day length (hours)	16	8
mean temperature (°C)	14	2

	summer (July)	winter (January)
day length (hours)	13	11
mean temperature (°C)	30	23

→ = migratory flyway

As the number of daylight hours decreases in late summer in Britain, hormonal changes occur within the birds' bodies that cause them to prepare for their autumn migration. The route taken is called a flyway. It is found to possess a wealth of prominent physical features such as coastlines, river valleys and mountain ranges.

Once the winter is over, the yellow wagtails leave Africa and return to Britain to breed. Each pair produces a clutch of 4–6 fertilised eggs which are incubated by the female for 2 weeks. Since yellow wagtails are diurnal birds, the parents spend the daylight hours searching for insect food to satisfy their hungry nestlings. After several weeks, the young become independent and when the summer draws to a close the cycle is repeated.

a) (i) During which season of the year do yellow wagtails migrate from Britain to West Africa?
(ii) Identify the environmental stimulus that triggers this form of rhythmical behaviour.
(iii) Suggest why the first frosty night following the summer season is not the trigger to which the birds' 'biological clock' responds.
b) Why do the birds migrate south via the coastal route rather than going more directly over the sea?
c) Name TWO climatic factors that could lead to the birds making navigational errors and losing their way.
d) With reference to the data in the diagram, explain the benefit to the yellow wagtail of spending the winter in West Africa.

e) Which word in the passage refers to a daily behaviour rhythm that means *'active during the day and inactive at night'*?
f) With reference to the data in the diagram, give ONE reason why the yellow wagtail benefits by returning to Britain for the summer.
g) Scientists studying migratory birds catch specimens and attach leg rings to them before releasing them.
(i) Explain the purpose of this procedure.
(ii) Suggest the information that would be carried by a leg ring.

The Body in Action

16 *Movement*

1 Match the terms in list X with their descriptions in list Y.

list X	list Y
1) antagonistic	**a)** framework of bones that supports and protects the body
2) ball and socket	**b)** smooth strong slippery material that reduces friction between bones
3) bone	**c)** tough inelastic material that attaches muscles to bones
4) cartilage	**d)** tough slightly elastic fibrous material that holds bones together
5) extensor	**e)** type of muscle that on contraction makes a limb bend
6) flexor	**f)** two muscles that produce movement of a limb in opposite directions
7) hinge	**g)** type of muscle that on contraction makes a limb straighten

8) ligament

9) skeleton

10) tendon

h) main component of the human skeleton consisting of living cells and hard mineral matter

i) type of joint that allows movement in three planes

j) type of joint that allows movement in one plane

2 The diagram below shows two positions of an athlete's right leg during a race. During the change from position 1 to position 2, which muscles in the diagram (i) contract (ii) become relaxed?

3 The human leg can be made to move in many different ways. Six of these movements are shown in the diagram opposite. Each movement, described in the following table, is brought about by a different group of muscles.
a) Match movements 1–6 with muscle groups A–F, which bring them about.

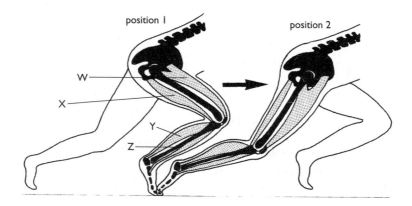

58

b) Each group of muscles in the table is antagonistic to another group also in the table. Identify the three pairs.

	before movement	after movement		before movement	after movement
1			4		
2			5		
3			6		

muscle group	description of movement brought about
A	two parts of limb pulled towards each other by closing a hinge joint
B	limb pulled forwards and out by movement at a ball and socket joint
C	limb pulled sideways and outwards away from body
D	limb pulled backwards and out by movement at ball and socket joint
E	two parts of limb pulled away from each other by opening a hinge joint
F	limb pulled sideways and inwards towards body

4 Read the following passage.

The long bones of a child are slightly pliable because they are still rich in cartilage; the long bones of older people contain little cartilage and are found to be more brittle.

If a bone is put under stress it may become fractured. In a closed fracture, the skin surface remains unbroken; in an open fracture, the skin is pierced by broken bone and bleeding is seen to occur. If, in addition, the broken ends of the bone cause injury to neighbouring blood vessels, nerves or organs, the fracture is described as complicated.

For satisfactory mending of a bone to take place, the broken ends must be correctly reunited and then immobilised. Healing begins immediately following the fracture. A blood clot forms between the broken ends of the bone. Within a few days the sharp surfaces of the broken bone are reabsorbed into the bloodstream leaving the ends soft and rubbery. These are held together by a bridge of fibrous tissue that is replaced by new bone detectable within 2–3 weeks using X-rays. By 6–12 weeks the process is well enough advanced to allow reuse of the affected limb.

a) Which type of fracture carries the higher risk of infection? Explain why.
b) Which TWO terms given in the passage would be used to describe the type of fracture illustrated in the first diagram below?

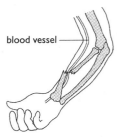

blood vessel

c) (i) The second diagram shows a greenstick fracture. In what way does this differ from the fractures described in the passage?

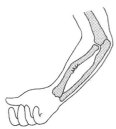

(ii) Greenstick fractures occur commonly amongst children but not amongst older people. Explain why.
d) (i) Which words in the passage mean 'having a broken bone set and the limb in plaster' for several weeks?
(ii) Predict the result of a broken bone not being properly set.

Problem Solving in Biology

e) Construct a flow chart to show the sequence of events that occurs during the bone-mending process.

5 a) A dry bone with mass of 50 g was placed in a metal dish of mass 20 g. After being heated for several hours to constant mass, the dish and contents were found to weigh 58 g. Calculate the percentage content of organic matter in the bone.
b) The table shown below refers to several chemical substances present in bone. Use the data in the table to identify the substances represented by bars X and Y in the accompanying bar chart.

chemical substance	% of body's total mass of substance present in bone
water	9
sodium	36
magnesium	50
phosphorus	88
calcium	99

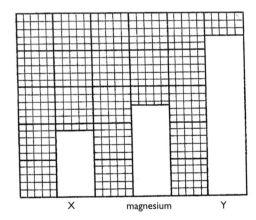

+

c) Following complete drying out, a bone was analysed and found to consist of 25% organic and 75% inorganic matter. 90% of its organic matter was made up of a structural protein called collagen. Present this information as a pie chart.

6 The following diagram shows a simple model of the human arm at rest and in two positions that it can adopt following muscular contraction.

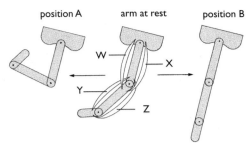

a) Identify the two muscles that must contract to allow the arm to adopt (i) position A (ii) position B.
b) Trace or copy the diagrams of positions A and B and then draw in all of the muscles in their relaxed or contracted state.
c) Which of muscles W, X, Y and Z are (i) extensors (ii) flexors?

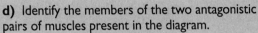
d) Identify the members of the two antagonistic pairs of muscles present in the diagram.

7 Which of the following graphs best represents the maximum load that can be supported at different arm extensions?

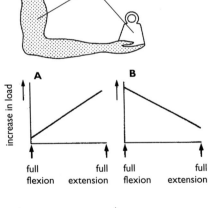

an arm supporting a load

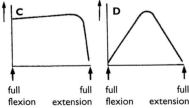

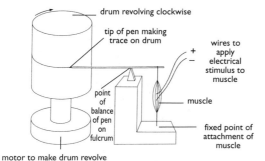

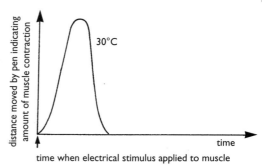

time when electrical stimulus applied to muscle

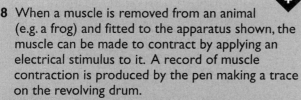

motor to make drum revolve

8 When a muscle is removed from an animal (e.g. a frog) and fitted to the apparatus shown, the muscle can be made to contract by applying an electrical stimulus to it. A record of muscle contraction is produced by the pen making a trace on the revolving drum.
a) In which direction will the tip of the pen move when the muscle (i) contracts, (ii) relaxes?
b) The two traces shown below were obtained by repeating the experiment at two different temperatures.
(i) Draw TWO conclusions from the results.
(ii) State TWO factors that must be kept constant during this experiment to allow a valid comparison to be made.

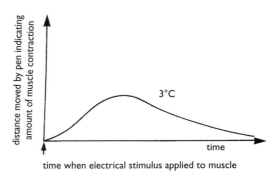

time when electrical stimulus applied to muscle

17 *The need for energy*

I Match the terms in list X with their descriptions in list Y.

list X

1) aorta

2) capillary

3) coronary artery

4) left atrium

5) left ventricle

list Y

a) structure present in heart and veins that prevents backflow of blood

b) upper chamber of heart that receives deoxygenated blood from the body

c) main vein that carries deoxygenated blood to the heart from the body

d) blood vessel that transports oxygenated blood from the lungs to the heart

e) first branch of the aorta that

6) pulmonary artery

7) pulmonary vein

8) right atrium

9) right ventricle

10) valve

11) vena cava

supplies oxygenated blood to the muscular wall of the heart

f) lower chamber of the heart that pumps blood to the lungs

g) main artery which carries oxygenated blood from the heart to the body

h) lower chamber of the heart that pumps oxygenated blood into the aorta

i) blood vessel that transports deoxygenated blood from the heart to the lungs

j) tiny thin-walled blood vessel in close contact with living cells

k) upper chamber of heart that receives oxygenated blood from the lungs

Problem Solving in Biology

2 The table below shows the amount of energy expended per hour by a 15-year-old girl when performing various activities.
Present the information as a bar chart.

activity	energy used (kJ/hour)
walking	1200
classwork	800
sport	2000
sitting at rest	330
sleeping	270

3 The volumes of air entering and leaving the lungs can be measured using a piece of apparatus in which a pen moves up and down on a revolving drum making a trace as the person breathes in and out. The following graph represents such a trace.

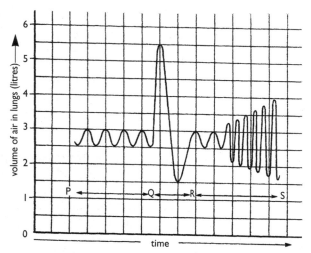

a) In which direction dose the pen move when air is being (i) inhaled (ii) exhaled?
b) What volume of air is inhaled and exhaled at each breath during P–Q when the subject is at rest?

c) Describe in detail the breathing pattern of the subject during period Q–R on the graph.
d) State the total lung capacity of this person as indicated by the graph.
e) State ONE factor that could account for the trace recorded during period R–S on the graph.

4 The following bar graph shows the rate of blood flow to various parts of a student's body under differing conditions of exercise.
a) Using this information, copy and complete the following table.

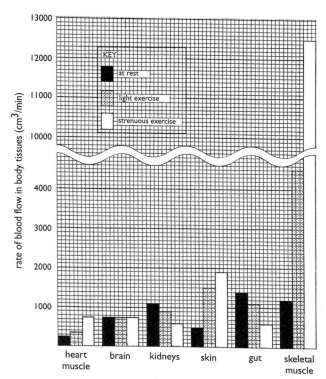

part of body	rate of blood flow (cm³/minute)		
	at rest		strenuous exercise
heart muscle		350	750
	750	750	
kidneys		900	600
	500		1900
gut	1400	1100	
skeletal muscle	1200		

b) What effect does increasingly strenuous exercise have on rate of blood flow to the (i) brain, (ii) skeletal muscle, (iii) gut?
c) Which other part(s) of the body shows the same trend in response to increase in exercise as the (i) gut, (ii) heart muscle?

d) Suggest a way in which the appearance of the skin would change as a result of strenuous, exercise. Explain why.

e) Calculate the **total** volume of blood per minute being pumped by the left ventricle to all of the parts of the body given in the graph during a period of (i) rest, (ii) light exercise, (iii) strenuous exercise.

f) If the student's pulse rate at rest is 65 beats/minute, what volume of blood is being pumped out of the left ventricle during each heart beat?

5 The accompanying diagrams show four simplified outlines of the human chest cavity. In which diagram do the arrows correctly indicate the movements of the diaphragm and rib cage during (i) inspiration, (ii) expiration?

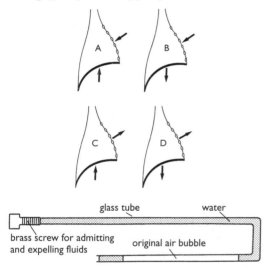

6 The above apparatus was used to investigate the composition of a bubble of exhaled air after exercise. With the bubble always at constant temperature and pressure, the following results were obtained:

length of gas bubble in contact with water = 100 mm

length of bubble after contact with potassium hydroxide (which removes CO_2) = 95 mm

length of bubble after contact with potassium pyrogallol (which removes oxygen) = 80 mm

a) Calculate the percentage of CO_2 present in the original bubble

b) Calculate the percentage of oxygen present in the original bubble.

7 Match the terms in list X with their descriptions in list Y.

list X	list Y
1) alveolus	a) sticky substance that traps dirt and is secreted by cells in the trachea
2) cartilage	b) dome-shaped muscular sheet at base of chest cavity involved in breathing movements
3) cilia	c) inhalation of air following contraction of intercostal muscles and diaphragm
4) diaphragm	d) tube connecting larynx with bronchi and allowing entry and exit of air
5) expiration	e) tiny air sac with thin lining that allows efficient gas exchange
6) inspiration	f) tough material found in incomplete rings that supports the trachea
7) intercostal muscle	g) fluid-filled space between chest wall and lungs that allows friction-free breathing movements
8) mucus	h) hair-like projections on inner surface of trachea that sweep dirty mucus upwards
9) pleural cavity	i) contractile component of chest wall involved in breathing movements
10) trachea	j) exhalation of air following relaxation of intercostal muscles and diaphragm

8 In the diagram shown overleaf, a capillary is in close contact with structure R. The relative concentrations of carbon dioxide (CO_2) and oxygen (O_2) are given at three different sites.

Problem Solving in Biology

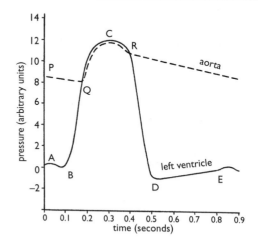

blood capillary

Which one of the following is correct?
A R is an alveolus and blood flow is from P to Q.
B R is an alveolus and blood flow is from Q to P.
C R is a muscle and blood flow is from P to Q.
D R is a muscle and blood flow is from Q to P.

9 The graph (right) shows the changes in pressure that occur in the left ventricle and the aorta during one heart beat.
a) What is the highest pressure recorded for the left ventricle?
b) Contraction of the left ventricle is represented by the part of the graph lettered: A–B, B–C, D–E, (choose the correct answer).
c) (i) At which lettered point on the graph does ventricular pressure first become equal to the pressure in the aorta? P, Q, R, (choose the correct answer).

(ii) After this point on the graph, for a short time ventricular pressure slightly exceeds aortic pressure. Which valve(s) will be open during this period?
(iii) At which lettered point on the graph will these valves start to close? P, Q, R, (choose correct answer).

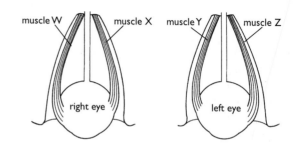

18 Co-ordination

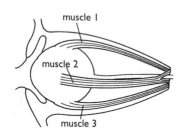

I The diagram (left) gives a view of the human eyeball from the side. Movements of the eyeball are controlled by six muscles, of which three are shown in the diagram.
a) Which muscle (attached to each eyeball) must contract to make the eyes perform movements (i), (ii), (iii) shown in the table?

	at start		after movement of eyeballs	
(i)				
(ii)				
(iii)				

The diagram below opposite gives a view of both eyeballs from above. Only two of the muscles controlling eyeball movements are shown for each eyeball. **b)** With reference to these muscles only, explain how the movement of the eyeballs shown below is brought about.

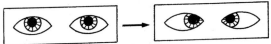

2 When a boy, sitting in a darkened room, had covered his left eye, a bright torch was switched on for 20 seconds and then the diameter of the pupil of his right eye was measured. This was called trial 1. The torch was switched off for 20 seconds and during this time moved to a position further away from the eye. It was then switched on again for 20 seconds and the diameter of the pupil measured (i.e. trial 2). This procedure was repeated for two more trials; and then four trials were done bringing the torch closer and closer to the eye each time.

The results are presented in the following bar chart.

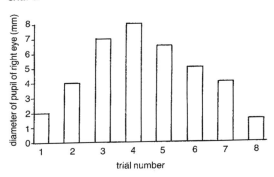

a) Between which two trials did the following changes in diameter of the pupil occur? (i) biggest increase, (ii) smallest increase, (iii) biggest decrease, (iv) smallest decrease.
b) At which trial was the torch (i) furthest away from the eye (ii) nearest to the eye?
c) At what two trials was the torch at the same distance from the eye?
d) In general, what relationship exists between the diameter of the pupil and the distance of the torch from the eye?

3 The accompanying diagram shows the field of vision of two types of animal: type A with eyes at the front of the head and type B with eyes at the side of the head.

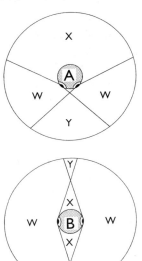

a) Which of letters W, X and Y represents: (i) the field of vision covered by both eyes; (ii) the field of vision covered by one eye only; (iii) the field covered by neither eye?
b) To which of animal types A and B would each of the following belong? (i) lion (ii) sparrow (iii) zebra (iv) hawk.

c) Relate the position of animal type A's eyes to its mode of feeding.
d) Relate the position of animal type B's eyes to its ability to survive in its natural environment.

4 Match the terms in list X with their descriptions in list Y.

list X	list Y
1) axon	**a)** region of brain responsible for memory, reasoning and imagination
2) central nervous system	**b)** region of brain responsible for control of rate of breathing and heart beat

3) cerebellum   **c)** sense organ that converts an environmental stimulus into a nerve impulse

4) cerebrum   **d)** arrangement of three different types of neurone through which an impulse passes resulting in a reflex action

5) effector   **e)** nerve cell in a reflex arc that transmits a nerve impulse from the receptor to the relay neurone

6) medulla   **f)** intermediate nerve cell in a reflex arc that transmits a nerve impulse from the sensory to the motor neurone

7) motor neurone   **g)** nerve cell in a reflex arc that transmits a nerve impulse from the relay neurone to the effector

8) receptor   **h)** fibre that carries a nerve impulse away from the cell body of a neurone

9) reflex action   **i)** structure that carries a nerve impulse towards the cell body of a neurone

10) reflex arc   **j)** tiny space between the axon ending of one neurone and the sensory fibre of the next neurone

11) relay neurone   **k)** muscle or gland that converts nerve impulse into a response

12) sensory fibre   **l)** region of the brain responsible for control of balance and muscular co-ordination

13) sensory neurone   **m)** part of the body composed of the brain and spinal cord

14) synapse   **n)** rapid automatic involuntary response to a stimulus

5 Each cerebral hemisphere of the human brain has a region called the sensory area and it is here that sensations such as touch and pain are perceived. Each part of the body capable of sending such impulses to the brain is represented by an area on the sensory region. However, the area of the brain devoted to each body part is found to be in proportion not to the actual size of the body part, but instead to the relative number of sensory receptors present in that body part.

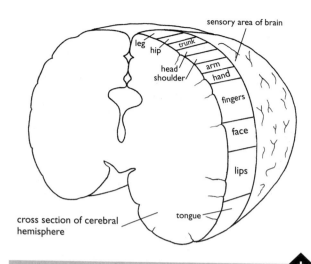

cross section of cerebral hemisphere

The diagram indicates how much of the brain's sensory area is given to each body part. The diagram below shows an imaginary human figure ('sensory homunculus') whose body parts have been drawn in relation to their sensitivity as opposed to their actual size.

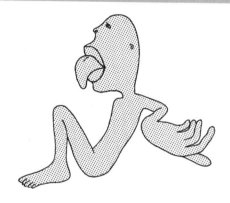

a) Name TWO types of sensation perceived by the sensory area of the cerebrum.
b) Account for the fact that a leg is a big part of a normal human body yet it is represented by a fairly small area on the cerebrum's sensory region.
c) Which body part contains most sense receptors relative to its actual size?
A trunk   **B** shoulder   **C** hip   **D** tongue
d) Which of the following structures has fewest nerve endings in relation to its actual size?
A arm   **B** face   **C** fingers   **D** lips

Each cerebral hemisphere also possesses a motor region that controls body movements. Here the size of the brain part allocated to each body part is related not to the body part's size but to its degree of mobility as illustrated by 'motor homunculus' shown below.

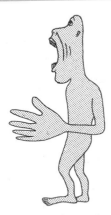

e) Identify TWO especially mobile parts of the body.
f) Which third part of the body is sufficiently mobile to be used, with much practice, to operate a pencil or paint brush?
g) With reference only to the pinna (ear flap), predict how 'motor homunculus' would differ if a rabbit had been drawn. Explain your answer.

6 A reflex action is a simple act of behaviour whose function is protective. In each case the reflex action is a rapid, automatic response to a stimulus. For example, when an object touches the eye the eyelid muscle contracts, bringing about blinking, which prevents damage to the eye. When foreign particles such as pepper enter the nasal tract, a sudden contraction of the chest muscles makes the person sneeze and remove the unwanted particles from the nose. Although reflex actions are involuntary, some can be partly altered by voluntary means; the person can to a certain extent resist blinking or sneezing. Some reflex actions cannot, however, be altered by voluntary means. When food is present in the gut, muscles in the gut wall contract bringing about peristalsis. This ensures efficient digestion by mixing the food thoroughly with digestive enzymes. In dim light the pupil of the eye becomes dilated (enlarged) as a result of movement of the iris muscle. This reflex action improves the person's vision in poor lighting. Peristalsis and pupil dilation cannot be resisted or prevented by voluntary means.

Copy and complete the table below using the information given in the passage.

reflex action	stimulus	response	protective function	can be partly altered by voluntary means?
blinking		contraction of eyelid muscle		yes
	presence of food in gut		ensures movement and therefore efficient digestion of food	no
	foreign particles in nasal tract	sudden contraction of chest muscles		
dilation of eye pupil			improves vision in poor lighting	

Problem Solving in Biology

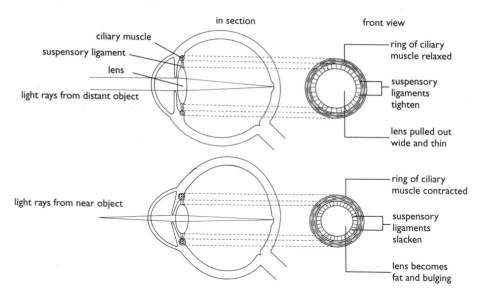

in section   front view

ciliary muscle
suspensory ligament
lens
light rays from distant object

- ring of ciliary muscle relaxed
- suspensory ligaments tighten
- lens pulled out wide and thin

light rays from near object

- ring of ciliary muscle contracted
- suspensory ligaments slacken
- lens becomes fat and bulging

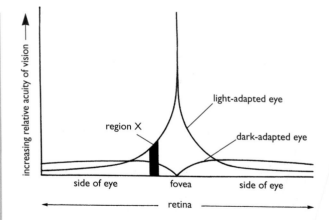

increasing relative acuity of vision →

region X

light-adapted eye

dark-adapted eye

side of eye   fovea   side of eye

← retina →

7 The lens of the human eye is surrounded by a ring of muscle fibres called ciliary muscle. This muscular ring is attached to the edge of the lens by threadlike suspensory ligaments as shown in the following diagrams.

In the upper eye the ring of ciliary muscle has become relaxed, increasing the diameter of the circle that it forms. This makes the suspensor ligaments become taut and pull the lens out into a wide, thin shape, which is ideal for bringing to a focus rays of light from a distant object.

To focus the rays of light coming from a near object, the shape of the lens has to be changed as shown in the lower diagram.

a) What shape has the lens become in order to focus light from a near object?

b) Describe in your own words how this change has been brought about.

8 When a person moves from bright light into almost total darkness, he is temporarily blinded, but after a few minutes the rod cells in each retina respond and he can see fairly well. His eyes are said to be dark-adapted. If he now returns to bright light he is briefly dazzled until the cone cells in each retina respond. Then he can see properly again and his eyes are said to be light-adapted.

The degree of sharpness of detail seen by an eye is called its visual acuity. This is compared for a dark-adapted and light-adapted eye in the following graph.

a) Explain the difference between a light-adapted and a dark-adapted eye.

b) (i) In which type of eye is the relative acuity of vision greatest at (1) the fovea, (2) the side of the eye?

(ii) Relate this difference to the distribution of rods and cones in the retina.

c) Acuity of vision for both types of eye is zero at region X on the retina. Suggest why.

d) A nocturnal bird was found to have a retina ⊕ composed entirely of rods. This means that it
A has good colour vision
B can judge distance well
C has accurate vision in dim light
D can only focus on distant objects
(choose the ONE correct answer)

9 The Eustachian tube is a structure about 40 mm long. It leads from the ear to a region of the throat (behind the nose) where it opens during yawning and swallowing. This allows air to pass into or out of the air-filled middle ear chamber. By this means, equal air pressure is maintained on either side of the ear drum, preventing it from bursting and allowing it to vibrate properly.

a) (i) Under what circumstances does the 'throat' end of the Eustachian tube open?
(ii) Why is this essential for the proper working of the ear?
b) What could happen to the eardrum if air pressure outside underwent a huge change but the Eustachian tube failed to open?
c) People often feel discomfort in their ears when travelling in a plane that is climbing to its cruising altitude. This is because air pressing on the outside of the eardrum is at a different pressure from that pressing on the inside. Why does sucking a sweet help to relieve the discomfort?
d) Suggest why it is possible to suffer temporary deafness during a very heavy cold when much catarrh is present in the nasal passages and throat.

19 *Changing levels of performance*

1 The table gives information about the average energy output of a British soldier when performing various activities before and after 3 months of training.
a) Construct a bar chart to illustrate all the information in the table.
b) Which activity was unaffected by training?
c) In general what effect did training have on all the other activities?

	energy output (kJ/min)	
activity	**before training**	**after training**
guard duty	10.9	10.7
ironing uniform for inspection	17.2	16.5
cleaning kit and rifle	11.3	11.3
assault course	41.2	35.7
slow marching	15.5	14.0
quick marching	22.7	20.3

2 The following graph refers to the pulse rates of two students X and Y. Student X trains regularly but Y does not undertake regular physical exercise.

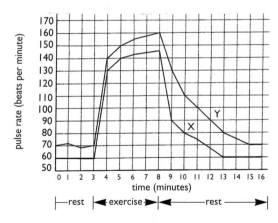

a) State the resting pulse rate for each student.
b) What was the pulse rate for each student 1 minute after exercise came to a halt?
c) From the graph, state TWO effects of training on pulse rate.

Problem Solving in Biology

3 The steps in the procedure used to measure length of recovery time after vigorous exercise are given below but in the wrong order.
(i) Exercise vigorously for 3 minutes (e.g. doing step test).
(ii) Calculate time required for pulse rate to return to normal.
(iii) Measure normal pulse rate before starting to exercise.
(iv) Take pulse rate at 1 minute intervals until normal pulse is recorded.
a) Arrange steps (i)–(iv) in the correct order.
b) One afternoon this procedure was used to compare the recovery times of several college students. Some students had been playing sports at lunchtime and had missed lunch. Some students had been studying until late the night before while others had gone to bed early.
(i) Explain why the investigation is not a fair test.
(ii) Suggest how it could be adapted to make it valid.

4 The following graph shows the effect of a period of exercise followed by a period of rest on the lactic acid concentration in the blood of a healthy, fit teenager.

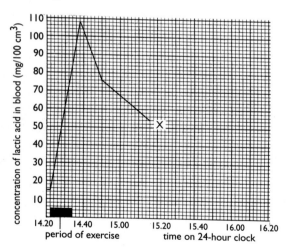

a) (i) At what time did the exercise stop? (ii) For how long did the period of exercise last?
b) (i) What was the highest concentration of lactic acid reached?
(ii) How long did it take for the concentration of lactic acid to drop from its highest level to 50% of its highest level?

c) (i) What was the concentration of lactic acid in the blood before the exercise period?
(ii) If the trend at X continues, at what time will the initial level of lactic acid in the blood be reached?

d) Suggest why the lactic acid concentration continued to rise for a brief time after the exercise stopped.
e) (i) Make a simple copy of the diagram and then add a line graph to represent the concentration of lactic acid for a super-fit athlete performing exactly the same exercise for the same length of time.
(ii) Explain the form taken by the graph that you have drawn as your answer to (i).

GRAPH A

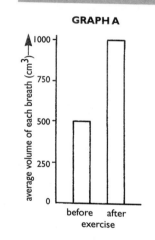

GRAPH B

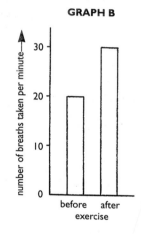

5 The above bar graphs show the effect of exercise on a girl's rate and depth of breathing.
a) Which graph shows (i) rate, (ii) depth of breathing?
b) What effect does exercise have on both rate and depth of breathing?

c) (i) What total volume of air passes in (and out) of the girl's lungs each minute after exercise?
(ii) By how much is this volume greater than that passing in (and out) of her lungs before exercise?
d) The air that the girl was breathing was found to contain 150 particles (pollen grains and fungal spores) per 500 cm³ of air.
(i) Calculate how many particles she would inhale per hour when not exercising.
(ii) Explain why very few of these particles would reach her alveoli.

70

6 Match the terms in list X with their descriptions in list Y.

list X	list Y
1) aerobic	**a)** state of muscle resulting from lack of oxygen and build-up of lactic acid
2) anaerobic	**b)** a measure of the number of times that the heart beats per minute
3) breathing rate	**c)** physical condition of body as indicated by length of recovery time
4) fatigue	**d)** chemical formed as a result of the incomplete breakdown of glucose during anaerobic respiration
5) fitness	**e)** type of respiration that occurs in the presence of oxygen and produces CO_2, water and much energy
6) lactic acid	**f)** course of vigorous exercise undertaken on a regular basis to improve fitness
7) oxygen debt	**g)** type of respiration that occurs in the absence of oxygen and produces lactic acid and a little energy
8) pulse rate	**h)** time taken for pulse and breathing rates to return to normal after a period of exercise
9) recovery time	**i)** state that develops during anaerobic respiration and is repaid during a rest period
10) training	**j)** a measure of the number of breaths taken per minute

7 The information in the following table at the bottom of the page refers to an alert, sober driver in a well-maintained car with good brakes and tyres.
a) Copy and complete the table.
b) In general, predict the effect of drinking four pints of beer on a driver's reaction time.
c) Describe how this would affect his overall stopping distance at any of the speeds given in the table.

8 The Harvard step test is often used to measure fitness. The subject steps onto and down from a 500 mm high platform 30 times per 5 minutes or until fatigue forces him to stop. His pulse rate is taken for the period exactly 1 to 1.5 minutes after exercise and his fitness index (FI) calculated using the formula:

$$FI = \frac{\text{total exercise time (in seconds)} \times 100}{5.5 \times \text{pulse rate}}$$

The following table shows the categories of fitness.

FI	fitness category
less than 50	poor
50–80	average
more than 80	good

The pulse rates of John, Joe and Jim were found to be 90, 120 and 65 respectively during the period 1 to 1.5 minutes after each had completed a 5 minute step test. To which fitness category does each boy belong?

car speed (miles/hour)		shortest stopping distance (in feet)		
		thinking distance	braking distance	overall stopping distance
20	B R A K E S A P P L I E D	20	20	40
30		30	45	75
40		40		120
50			125	175
60		60		
70		70	245	315

6 Inheritance

20 Variation

1 Match the terms in list X with their descriptions in list Y.

list X

1) bar chart (graph)

2) continuous

3) discontinuous

4) fertile offspring

5) histogram

6) normal distribution

7) species

8) sterile offspring

list Y

a) group of interbreeding organisms whose offspring are fertile

b) young that are able to produce more of their own kind on reaching sexual maturity

c) young that are unable to produce more of their own kind on reaching sexual maturity

d) block graph of a characteristic showing continuous variation where the entire range of the characteristic is divided into small groups for convenience

e) type of variation shown by a characteristic when it varies in an uninterrupted way from one extreme to another

f) type of variation shown by a characteristic that can be used to divide up the members of a species into two or more distinct groups

g) graph consisting of separate blocks each of which represents a distinct category of a characteristic showing discontinuous variation

h) bell-shaped curve of a characteristic showing continuous variation where most individuals fall into the centre of the range

2 Which of the following is an example of discontinuous variation amongst humans?
A waist circumference
B male or female sex
C length of handspan
D breadth of foot
(Choose the ONE correct answer.)

3 Which of the following is an example of continuous variation amongst humans?
A intelligence
B ear lobe type
C wavy or straight hair
D right or left-handedness
(Choose the ONE correct answer.)

4 After a visit to a zoo, 2000 young schoolchildren were asked to name the species of newly born animal that they liked the most. The results of the survey are presented in the following bar graph.

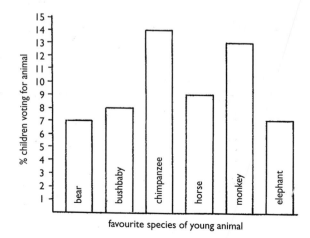

favourite species of young animal

a) By how many percent was the most popular species of animal the clear favourite over its nearest rival?
b) What total percentage of the children voted for the chimpanzee, the monkey and the bushbaby?
c) What percentage of the children voted for other animals not shown in the bar graph?
d) In reply to the question 'Which species of young animal do you dislike the most?', 540 children out of the 2000 chose the baby snake. Express this number of children as a percentage of the total.
e) It was found that 8% of the children decided that the tarantula spider was their least favourite baby animal. How many children held this view?

5 Draw up a table with the name of each of the following nine people in the extreme left hand column. Complete the table to show five differences given in the diagram that exist between the nine people.

6 The following table shows the percentage distribution of blood groups in the United Kingdom.

	blood group type (%)			
	A	B	AB	O
Scottish	34	11	3	52
English	42	8	3	
Irish	26		7	32
Welsh	38	10		49

a) Present the Scottish figures only, as a bar graph.
b) Copy and complete the table.
c) Calculate the UK average percentage for each blood group and extend your table to include these figures.
d) Is blood group an example of continuous or discontinuous variation?

7 The following histogram shows the variation in mass of the individual grapes that made up a bunch.
a) Describe the distribution of grape mass in terms of range and most common value for this population.
b) How many grapes were found to have the most common value?
c) What was the least common mass in this survey?
d) How many grapes had a mass of 2.8 g?
e) How many grapes were weighed?
f) What percentage of grapes had a mass of 2.5 g?
g) How many grapes weighed more than 2.7 g?
h) Calculate the average mass of a grape for this population.

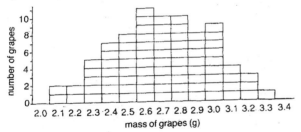

8 A species is a group of organisms whose members are able to interbreed and produce fertile offspring.

  The table shows the results of crosses between eight different types of animal. A shaded box means that fertile offspring resulted from the cross. A blank box means that the cross failed (i.e. sterile or no offspring produced).
How many different species of animal were present?

9 The following list gives the steps involved in constructing a histogram to show the variation in mass that exists amongst the fish in a fish farm. Arrange the five steps into the correct sequence.
a) Consider the range in mass obtained with respect to the lightest and heaviest.
b) Prepare graph paper by putting range in mass as sub-sets on the x-axis and number of fish on the y-axis.
c) Weigh a large sample e.g. 2000 fish.
d) Count the number of fish in each sub-set and draw the histogram.
e) Divide the range in mass into several (e.g. 10) equally spaced sub-sets.

10 The lengths of 60 leaves from an apple tree are listed in the table opposite in random order. Divide the group up into nine sub-sets differing from each other by 10 cm (with the first sub-set being 21–30 cm) and then present the information as a histogram.

animal type

leaf no.	length (cm)	leaf no.	length (cm)	leaf no.	length (cm)
1	48	21	90	41	92
2	31	22	97	42	63
3	52	23	66	43	69
4	61	24	59	44	69
5	57	25	56	45	60
6	77	26	64	46	57
7	86	27	62	47	43
8	81	28	75	48	67
9	103	29	83	49	70
10	72	30	99	50	80
11	80	31	33	51	84
12	68	32	45	52	47
13	65	33	28	53	79
14	58	34	42	54	76
15	45	35	56	55	95
16	39	36	61	56	58
17	62	37	64	57	55
18	61	38	77	58	74
19	76	39	89	59	54
20	78	40	81	60	78

21 *What is inheritance?*

1 In a genetics experiment, a pupil crossed six ebony-bodied male fruit flies with one wild type female in a culture tube. While the tube was in the incubator at 25°C, the female laid 59 eggs. After 10 days the pupil inspected the tube and found it to contain 60 wild type flies and six ebony-bodied. She concluded therefore that these flies made up the F_1 generation.
a) Spot an error in the girl's experiment that accounts for the strange result.
b) Assume that you have been asked to repeat this experiment. In addition to correcting the girl's procedural error, name ONE further change that you would make to improve the experiment.

2 Read the following passage.

In 1866 Gregor Mendel, an Austrian monk, published the results and conclusions from breeding experiments on pea plants. His work is now regarded as brilliant for its time.

  He understood the importance of studying only one difference between parent plants at a time. Not content with describing the offspring produced, he also carefully counted them making it possible to express his results quantitatively (e.g. as a ratio).

  Without knowing about chromosomes or genes, he was able to interpret his results

and from them make up laws of genetics that still stand to this day. However, during his lifetime, Mendel was not regarded as an important scientist and it was not until 1900, 16 years after his death, that his work was found by a Dutch biologist and recognised for its true worth.

a) Choose the best title for the above passage from the following:
A Sexual reproduction in pea plants
B Mendel, the father of modern day genetics
C Famous scientists of the 19th century
D Why Mendel's work remained ignored.
b) Give TWO examples of sound scientific procedure followed by Mendel in his experimental work.
c) Suggest why Mendel is regarded as 'a man ahead of his time'.

3 Copy and complete the accompanying diagram of the human life cycle, using the following terms to fill the blank boxes: egg, growth, zygote, gamete formation.

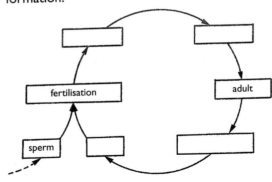

4 The following diagrams show the chromosomes present in the nuclei of various types of cell from the fruit fly, *Drosophila*.

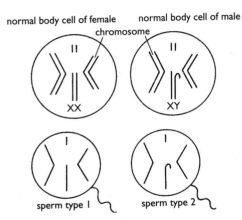

a) (i) Which of the following diagrams represents the set of chromosomes that would be found in a zygote formed as a result of sperm type 1 fertilising a normal egg?
(ii) State which sex this zygote would be.
b) (i) Which of the following diagrams represents the set of chromosomes that would be found in a zygote formed as a result of sperm type 2 fertilising a normal egg?
(ii) State which sex this zygote would be.

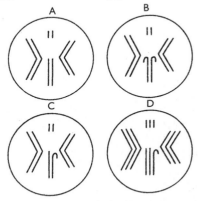

5 In pea plants the gene for height has two gene forms (alleles), tall and dwarf. The following cross was carried out:

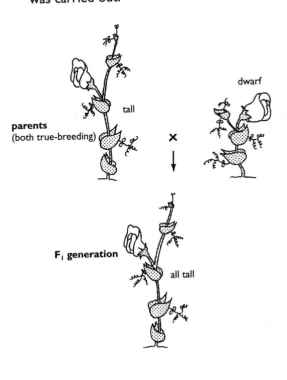

Problem Solving in Biology

a) Which gene form (allele) is dominant? Explain your answer.

b) Using letters of your own choice, give the genotype(s) of the F_1 generation.

c) A further generation of pea plants was produced by allowing the F_1 generation to self-pollinate. Using your chosen symbols, copy and complete the following table to show the outcome of this cross.

	genotypes of pollen	
genotypes of ovules		

d) The F_2 generation consisted of 840 plants. In theory, how many would be (i) homozygous for tallness, (ii) heterozygous for tallness, (iii) dwarf?

6 Match the words in list X with their descriptions in list Y.

list X

1) alleles
2) chromosome
3) dominant
4) gametes
5) gene
6) genotype
7) heterozygous
8) homozygous

list Y

a) type of sex chromosome present in human males and females

b) sex cells that carry genes from one generation to the next

c) type of sex chromosome present in human males only

d) basic unit of inheritance many of which make up a chromosome; each controls an inherited characteristic

e) one of several thread-like structures found inside the nucleus of a cell

f) alternative forms of a gene that are responsible for different expressions of an inherited characteristic

g) describing the member of a pair of alleles that always shows its effect and masks the presence of the recessive allele

h) describing the member of a pair of alleles that is always masked by the dominant allele

9) phenotype
10) recessive
11) X chromosome
12) Y chromosome

i) the set of genes possessed by an organism

j) physical appearance of an organism

k) describing a genotype that contains two identical alleles of a particular gene

l) describing a genotype that contains two different alleles of a particular gene

7 In humans, wavy hair is dominant to straight hair. A wavy-haired woman marries a straight-haired man and they have four children. One son and one daughter both have wavy hair and one son and one daughter both have straight hair.

a) Present this information as a family tree, giving a key to all the symbols that you use.

b) Using letters of your choice, label the tree with the genotypes of (i) each parent, (ii) each child.

c) Explain how you arrived at your answer to part **b)** (ii).

8 The results displayed in the table below were obtained by first crossing a true-breeding black mouse with a brown mouse. (Black is dominant to brown.) The members of the F_1 generation consisted of three males and five females. These were used in turn as the parents of the F_2 generation.

a) Total up the F_2 and express it as a phenotypic ratio.

b) Suggest why it is better to consider 10 F_2 litters instead of just one.

	number of black mice	number of black mice
original parents	1	1
F_1	8	0
F_2 litter 1	5	3
litter 2	5	3
litter 3	7	0
litter 4	7	2
litter 5	5	2
litter 6	6	0
litter 7	5	4
litter 8	7	2
litter 9	7	2
litter 10	6	2

9 A pupil was asked to design an experiment to show that crossing male wild type fruit flies with female dumpy-winged flies is genetically the same as crossing male dumpy-winged flies with female wild type flies. He was given the apparatus shown below (where all the flies are true-breeding).

The following diagram shows what he proposed to set up as his test.

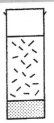 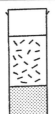

A containing 6 ♂ wild type and 14 ♀ dumpy-winged

B containing 17 ♂ dumpy-winged and 3 ♀ wild type

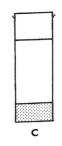

C   D

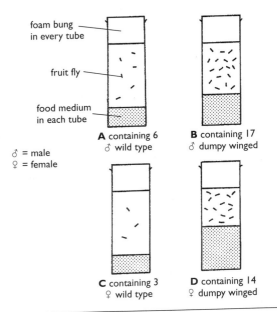

foam bung in every tube

fruit fly

food medium in each tube

A containing 6 ♂ wild type

B containing 17 ♂ dumpy winged

♂ = male
♀ = female

C containing 3 ♀ wild type

D containing 14 ♀ dumpy winged

State TWO ways in which his experimental set-up would have to be altered to make it fair and allow a valid comparison to be made.

---

10 In each of the following crosses, the original parents were true-breeding and the members of each F₁ generation were crossed with one another.

Example 1:
Mouse fur colour where black is dominant to brown

	number of black	number of brown
parents	1	1
F₁	9	0
F₂	60	20

Example 2:
Guinea-pig hair type where straight is dominant to wavy

	number of straight	number of wavy
parents	1	1
F₁	7	0
F₂	29	11

Example 3:
Hamster fur colour where golden is dominant to white

	number of golden	number of white
parents	1	1
F₁	8	0
F₂	45	16

a) What generalisation about the phenotypic ratio of the F$_2$ generation can be drawn from this information?

b) In gerbils brown coat colour is dominant to grey. Copy and complete the following table where a true-breeding brown gerbil is crossed with a grey mate and their offspring are then crossed with one another, by inserting the expected numbers.

	number of brown	number of grey
parents	1	1
F$_1$	6	
F$_2$		8

11 In pea plants, the allele for flower colour (**C**) is dominant to the allele for lack of flower colour (**c**). A plant homozygous for flower colour was crossed with a plant bearing colourless flowers. The F$_1$ plants were then self-pollinated.

Which of the following correctly represents the ratio of genotypes expected in the F$_2$ generation?

(i) all **Cc**   (ii) 1**CC**:1**Cc**   (iii) 3**CC**:1**cc**
(iv) 1**CC**:2**Cc**:1**cc**

---

22 Genetics and society

1 Read the following passage.

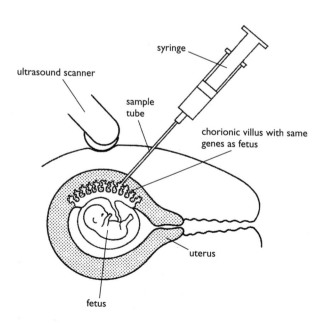

- ultrasound scanner
- syringe
- sample tube
- chorionic villus with same genes as fetus
- uterus
- fetus

A technique for diagnosing genetic diseases in unborn children is known as chorionic villus (**CV**) sampling. A tiny sample of tissue from the placenta is removed from the mother's womb using a special syringe guided by ultrasound scanners as shown in the accompanying diagram.

The cells (genetically identical to the fetus) are inspected in the laboratory to see if their genetic material has unusual characteristics that indicate that the baby will suffer from an incurable disease. If this proves to be the case, the parents can decide to have the pregnancy terminated. This technique can be done as early as 8 weeks into the pregnancy when an abortion may be less distressful to the mother than at 18–20 weeks (which is when the alternative test, amniocentesis, is normally done). However, initial studies have shown that **CV** sampling is followed by a miscarriage in about 3 out of every 100 of those mothers whose babies are found to be normal.

a) By what means can a sample of tissue, genetically identical to an unborn baby, be obtained without touching the baby?
b) Why is this tissue inspected in the laboratory?
c) Suggest another name for the 'genetic material' referred to in lines 8 to 9.
d) Give ONE way in which the above technique is preferable to amniocentesis.
e) State a disadvantage of CV testing.

2 The following diagram shows an aurochs (the extinct ancestor of modern cattle) and a British Friesian cow (one of many modern breeds of cattle).

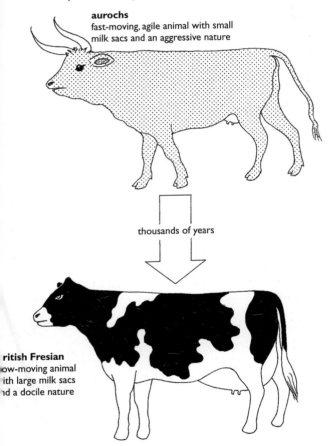

aurochs
fast-moving, agile animal with small milk sacs and an aggressive nature

thousands of years

British Fresian
slow-moving animal with large milk sacs and a docile nature

a) Describe the method employed by humans to bring about the changes indicated in the diagram.
b) Give THREE reasons why a modern farmer would prefer a herd of British Friesian cattle to a herd of aurochs.
c) What possible use could be made of the aurochs nowadays if it had not been allowed to become extinct?

3 Read the following passage.

In 1791, a sheep farmer discovered a completely new variety amongst his flock. A male lamb had been born with short legs. We now know that this new form of leg length had arisen as a result of the lamb inheriting a changed gene that neither parent possessed.

The farmer decided that he would like a whole flock of short-legged sheep since the low fences needed to keep them in the fields would save him money on materials. He therefore crossed the unusual ram with a normal ewe as shown in the diagram.

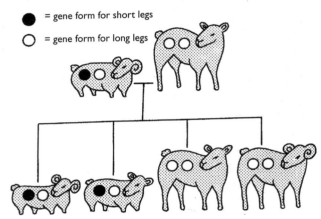

● = gene form for short legs
○ = gene form for long legs

He then crossed only the short-legged animals with one another for several generations and eventually produced a flock of short-legged sheep.

a) Which words in the passage mean a 'mutation'?
b) Is the gene form (allele) for short legs dominant or recessive? Explain your answer with reference to the diagram.
c) Using the circular symbols, make a keyed diagram to show the outcome on average of crossing the two short-legged animals shown in the diagram. State the phenotype of each offspring produced.
d) (i) Is short-leggedness in sheep an example of advantageous or disadvantageous variation to the farmer? Explain your answer. **(ii)** Would short-leggedness in sheep be advantageous or disadvantageous to wild sheep before domestication by man? Explain your answer.
e) Which sentence in the passage refers to an example of selective breeding over a long period of time?

Problem Solving in Biology

4 The following table refers to six breeds of sheep that have been produced by selectively breeding them for their wool.

breed	average fleece weight (kg)	colour of face	average length of wool (mm)	main use of wool
Cheviot	2	white	100	cloth for suits
Scottish Blackface	2.5	black	260	carpets
Shetland	1.1	white	100	knitwear
Border Leicester	3.6	white	200	knitwear
Exmoor Horn	2.7	white	90	cloth for suits
Rough Fell	2.3	black and white	200	carpets

a) Give THREE characteristics of Rough Fell sheep.
b) Which TWO breeds of sheep have a white face and wool used for making suits?
c) A sheep was found to have a fleece weighing 3.7 kg and white wool of length 195 mm that was suitable for knitting. Identify its breed.

d) Copy and complete the following key of paired statements using the information given in the table.

```
1  face not completely white . . . . . . . . . . . .go to 2
   face completely white  . . . . . . . . . . . . . .go to 3
2  _____ . . . . . . . . . . . . . . . . . . . _____
   _____ . . . . . . . . . . . . . . . . . . . _____
3  fleece weighs less than 2.5 kg on average  .go to 4
   fleece weighs 2.5 kg or more on average  .go to 5
4  _____ . . . . . . . . . . . . . . . . . . . _____
   _____ . . . . . . . . . . . . . . . . . . . _____
5  average length of wool less than 100 m.
       . . . . . . . . . . . . . . . . . . . . . . . . . .Exmoor Horn
```

5 The table below shows the results of an investigation into the effect of increasing doses of radiation on germinating barley grains, which were planted in batches of one hundred.

units of radiation (roentgens)	0	1000	2000	3000	4000	5000
number of barley grains germinating and producing a shoot	96	32	16	10	6	2
number of shoots with abnormal leaves	0	16	12	9	6	2
% number of shoots with abnormal leaves						

a) Construct a line graph to show the effect of increasing radiation on germination of barley grains. (Put units of radiation on the x-axis.)
b) Draw a conclusion from your graph.
c) Explain why batches containing as many as one hundred grains were planted.

d) Calculate the percentage number of shoots with abnormal leaves for each dose of radiation (as indicated by the blank boxes in the table).
e) Add a second vertical scale to the right hand side of your graph and plot the effect of increasing radiation on the percentage number of shoots with abnormal leaves as a second line graph.
f) Draw a conclusion from your second line graph.
g) Predict the dosage of radiation that would cause 80% of the barley grains to fail to germinate.

6 Cystic fibrosis is an inherited disorder of the human body. It affects mucus production, causing blockage of tiny air passages in the lungs. It is due to a recessive gene form.
  Consider the following family trees:

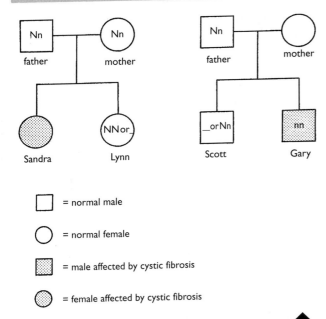

☐ = normal male

◯ = normal female

▨ = male affected by cystic fibrosis

⊙ = female affected by cystic fibrosis

a) Copy and complete the diagram by inserting the missing genotypes.
  Lynn and Scott are intending to marry so they seek the advice of a genetic counsellor. The counsellor studies their family trees and figures out that if the couple marry then this would result in one of the following crosses:

(i) NN × NN   (ii) NN × Nn   (iii) Nn × Nn

b) Which TWO crosses involve no risk of producing affected children?
c) In the remaining cross, what is the chance of each child being affected?
d) Show in diagrammatic form how you arrived at your answer to **c)**.

7 The graph shown below refers to an investigation into the effect of increasing radiation on the percentage number of X chromosomes showing a lethal (deadly) mutation.
  The animal used was the fruit fly (*Drosophila melanogaster*).
  The results are plotted as points with the best straight line drawn through them.

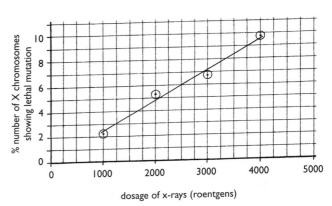

dosage of x-rays (roentgens)

a) What was the ONE variable factor altered by the experimenter in this investigation?
b) How many conditions of this factor were used?
c) Name TWO other factors that would have to be kept constant to make the experiment valid.
d) What conclusion can be drawn from the above results?

Biotechnology

23 Living factories

1 Match the terms in list X with their descriptions in list Y.

list X

1) alcohol
2) bacteria
3) biotechnology
4) bloom
5) curdling
6) fermentation
7) lactic acid
8) pasteurisation
9) yeast
10) yoghurt

list Y

a) single-celled fungus used in brewing and baking

b) energy-releasing process carried out by many micro-organisms in the absence of oxygen

c) single-celled organisms used to make cheese and yoghurt

d) substance produced by yeast during fermentation of glucose

e) substance produced by bacteria when milk turns sour

f) food produced from milk by action of lactic acid-forming bacteria

g) 'dust' containing wild yeast on the surface of fruit

h) coagulation of milk caused by the action of lactic acid

i) employment of living cells to produce substances that are useful to human beings

j) heating of milk to 73°C for 15 seconds to kill harmful micro-organisms

2 The experiment shown in the accompanying diagram was set up to investigate the effect of live yeast on glucose solution. It was left in an incubator at 30°C for 2 hours.

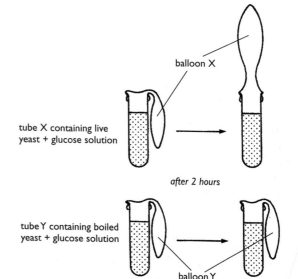

tube X containing live yeast + glucose solution

balloon X

after 2 hours

tube Y containing boiled yeast + glucose solution

balloon Y

a) Explain why balloon X has inflated but balloon Y has failed to do so.

b) Predict the effect of repeating the experiment at (i) 10°C and (ii) 70°C on the time required to inflate balloon X to the size shown in the diagram. Explain your answer in each case.

c) With reference to the glucose solution, identify a feature not shown in the diagram that must be kept constant in tubes X and Y for a valid comparison of the results to be made.

3 The following table refers to six types of bacteria.

a) Which type of bacterium fails to become purple after gram staining?

b) Consider the sphere-shaped bacterial types in the table. Apart from shape and becoming purple after gram staining, what TWO other features do they have in common?

scientific name of bacterium	shape	motile?	colour after gram stain	spore-forming?
Lactobacillus	rod	−	purple	−
Clostridium	rod	+	purple	+
Streptococcus	sphere	−	purple	−
Salmonella	rod	+	pink	−
Staphylococcus	sphere	−	purple	−
Bacillus	rod	+	purple	+

(+ = yes, − = no)

c) Consider the spore-forming types of bacteria and identify THREE other features that they have in common.
d) Which type of bacterium is rod-shaped, purple after staining but fails to make spores?
e) How many types of bacteria are motile and spore-forming?
f) Identify TWO different pairs of bacterial types that each have all four features in common.

4 The experiment shown in the diagram below was set up to investigate whether live yeast produces heat when respiring anaerobically (i.e. in the total absence of oxygen). State TWO ways in which the experiment will have to be altered in order to make it a valid test.

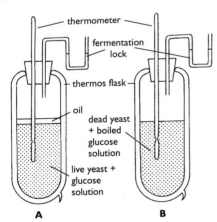

A   **B**

5 Sugar is broken down into lactic acid by a type of bacterium called *Lactobacillus*. Calcium carbonate is a white insoluble solid that reacts with acid to form a clear colourless solution.

A plate of agar containing sugar and calcium carbonate (which gives agar a cloudy appearance) was prepared and then four different types of bacteria (A, B, C and D) were added to it as shown in the following diagram.

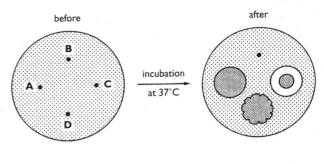

= bacterial growth   = clear agar   = cloudy agar

a) Identify *Lactobacillus*.
b) Give a reason for your choice.

6 The rate of carbon dioxide production from a yeast culture was measured at regular intervals over a period of 60 hours and the results tabulated as follows.

time (hours)	rate of CO_2 production (cm³/hour)
0	0
10	30
20	60
30	90
40	110
50	116
60	116

a) Using graph paper similar to that shown, and the axes (which have been partly completed for you), present the above results as a line graph.

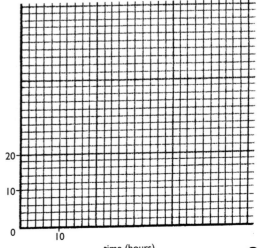

time (hours)

Problem Solving in Biology

b) How long did it take until CO_2 was being produced at a rate of 84 cm^3/hour?

c) State the volume of CO_2 being produced per hour at (i) 14 hours after the start, (ii) 56 hours after the start.

7 Natural yoghurt contains live bacteria that convert sterile milk into yoghurt. A boy was asked to set up an experiment to find out the best (optimum) temperature for yoghurt formation. He rinsed three glass beakers with hot water and dried them on clean paper towels. He then used them to set up the following experiment.

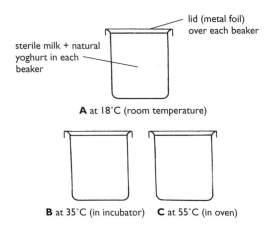

A at 18°C (room temperature)

B at 35°C (in incubator)   **C** at 55°C (in oven)

Results (after 6 hours)

	texture of 'yoghurt'	smell of 'yoghurt'
A	watery	faintly unpleasant
B	creamy	unpleasant
C	watery	no smell

a) Spot the important error in the boy's procedure that probably led to the unpleasant smell developing.

The error was corrected in a second attempt at the experiment and the smell did not develop. This time the results were as follows.

	texture of 'yoghurt'
A	watery
B	creamy
C	watery

The boy concluded that 35°C is the best temperature for yoghurt formation.
b) Suggest how his experiment could be further improved to find out a more accurate estimate of the optimum temperature.

8 Two groups of pupils set up an experiment to investigate the effect of temperature on the action of baker's yeast. Group A made their dough from flour, yeast and water; group B used flour, yeast, sugar and water. Each poured dough into three measuring cylinders up to the 10 cm^3 mark and left the cylinders at 5°C, 20°C and 35°C. The following graphs on p.85 shows their results.

a) (i) Which variable factor did both groups set out to investigate?
(ii) In both cases what overall effect did this factor have on the action of baker's yeast?
b) (i) By what ONE factor did group A's investigation differ from that of B at the start?
(ii) What overall effect did this factor have on the action of baker's yeast?
(iii) Explain your answer to (ii).
c) By how many cm^3 was the volume of group B's dough greater than that of group A at 30 minutes and 20°C?

d) By how many times was the volume of group B's dough greater than that of group A at 50 minutes and 35°C?
e) (i) Predict the result of repeating the experiments at 250°C (the temperature at which bread is baked). (ii) Explain your answer to (i).

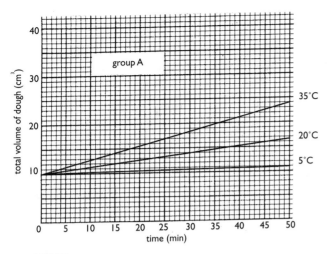

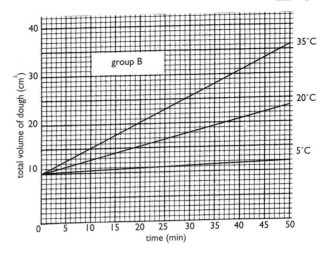

24 *Problems and profit with waste*

1 Match the terms in list X with their descriptions in list Y.

list X

1) activated sludge process

2) autoclave

3) biological filtration

4) contaminant

5) decay

6) disinfectant

7) dysentery

8) inoculation

9) methane

10) mycoprotein

11) sewage

12) single-celled protein

13) sterile

14) upgrading

list Y

a) organic waste material containing faeces, soap and food fragments

b) a water-borne disease spread by untreated sewage

c) substance used to kill micro-organisms outside the human body

d) completely free of micro-organisms

e) conversion of waste to useful products of higher economic value

f) substance produced by fungi that can be made into meat-like products

g) apparatus used to sterilise glassware and chemicals

h) useful gas produced by bacterial fermentation of sewage

i) unwanted micro-organism that has gained access to a culture or food medium

j) substance composed of bacterial cells that can be used to feed farm animals

k) introduction of a micro-organism to an environment containing the requirements for its growth

l) method of sewage treatment involving decay bacteria and a bed of stones containing air spaces

m) process by which decomposers obtain their energy by breaking down organic matter

n) method of sewage treatment involving the addition of decay bacteria and compressed air

2 A pupil was asked to find out if sour milk contains more bacteria than fresh milk. The series of diagrams shows the procedure that he followed.
Spot TWO mistakes that the boy made and then copy and complete the table at the foot of the page.

error or omission in procedure	correct procedure	reason for following correct procedure

Problem Solving in Biology

1 He collected the following apparatus

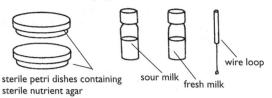

sterile petri dishes containing
sterile nutrient agar

sour milk

fresh milk

wire loop

2 He held the wire loop in a Bunsen flame until it was red hot and then cooled it by waving it in the air

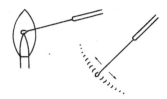

3 He collected a small drop of sour milk on the loop

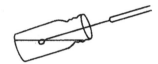

4 He spread the sour milk over the nutrient agar in one of the plates

5 He then used the wire loop to take a sample of fresh milk which he spread in the same way onto the second plate of agar

6 He sealed the dishes with sticky tape and put them in a warm incubator

3 **The rate at which an organism increases in dry weight is often taken as an indication of growth. The maximum percentage dry weight increase shown by different organisms (all grown under ideal conditions for 24 hours) varies enormously. Large multicellular animals grow much more slowly than tiny unicellular organisms. The foetus of a cow, for example, only increases by 1% of its dry weight over 24 hours whereas yeast cells increase by 1400% and bacteria by 4750% of their dry weight over the same period. When compared with microbes, multicellular plants similarly show a modest percentage dry weight increase. A Scots pine tree only gains 5% and even a fast growing plant such as sunflower 29% in 24 hours.**

Present the information contained in the above passage in a table so that the data can be easily compared.

4 Read the following passage.

Single-celled protein (SCP) is produced by certain strains of unicellular microbes cultured on petroleum hydrocarbons. A large mass of this protein, which is used for animal feed, can be produced quickly, since single-celled micro-organisms grow at a very rapid rate. The protein content of the product is high and its production is independent of climate. However, SCP contains a high concentration of nucleic acid. If eaten by humans this would be converted to uric acid which would gather as crystals in the joints causing a painful condition called gout. In farm animals, uric acid is converted by an enzyme into a soluble waste, which is later excreted.

Mycoprotein is produced by a fungus called _Fusarium_ growing on glucose syrup. Although able to double its weight every 5 hours, this multicellular fungus grows more slowly than single-celled microbes. Its product, mycoprotein, does not contain a high concentration of nucleic acid. Unlike unicellular micro-organisms, _Fusarium_ has a thread-like structure resembling meat fibres in size and strength. It can therefore be spun into a meat-like substance by interlaying the fibres with suitable colouring and flavouring. Food technologists have already used mycoprotein to produce meatless 'burgers' and 'sausages' that contain 44% protein and do not shrink on cooking. By varying the flavouring and other additives, they are now producing realistic 'chicken and ham paté', 'game pie' and even 'chocolate biscuits'. The manufacturers, concerned that some consumers might be put off these foods by

the thought of eating fungus, are quick to point out that mushrooms, a traditional and widely accepted food, are also fungi.

a) Apart from high protein content, what two advantages are gained by producing SCP for animal feed rather than using a plant crop such as turnips?

b) Explain why SCP can be fed to farm animals yet is unsuitable for human consumption.

c) Unlike SCP, mycoprotein is safe for humans to eat. Explain why.

d) (i) Describe the texture of the fungus *Fusarium*.
(ii) Of what advantage is this texture to food technologists?

e) (i) Why might foodstuffs made of mycoprotein meet with consumer resistance?
(ii) What defence to such criticism is offered by the manufacturers?

5 The following experiment was set up to investigate the production of methane by microbes respiring in the absence of oxygen. It has been running for several days.

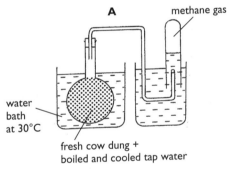

A   methane gas

water
bath
at 30°C

fresh cow dung +
boiled and cooled tap water

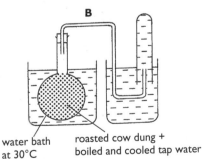

B

water bath
at 30°C

roasted cow dung +
boiled and cooled tap water

a) At the start, test tube A was full of water but now it contains some methane gas. Account for the fact that no methane is present in tube B.

b) Give TWO possible reasons for boiling the tap water before use in the above experiment.

6 When grown in a bioreactor at 30°C, the mycoprotein fungus doubles its weight every 5 hours as shown in the following partly completed table of results.

time (hours)	mass of fungus (g)
0	100
5	
10	400
	1600

a) Copy and complete the table.

b) Draw a line graph of this information.

c) Predict what would happen to the growth rate of the fungus at 10°C.

d) On your graph paper, draw a further line that could represent the growth rate of the fungus at 10°C.

7 The graph below shows the number of bacteria growing in nutrient broth kept at constant optimum temperature over a period of 30 hours.

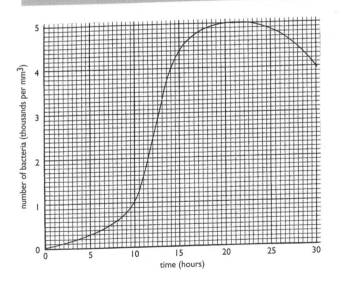

Problem Solving in Biology

bottle	treatment		
	heated in pressure cooker at 121°C for 20 minutes	pasteurised at 73°C for 15 seconds	place where bottle was then kept for 1 week
W	no	yes	warm room
X	yes	no	fridge
Y	no	no	warm room
Z	no	yes	fridge

a) How many bacteria were present per mm³ at 5 hours?
b) From this point in time onwards, how many more hours passed before the bacteria had doubled in number?
c) During which of the following periods of time was growth rate greatest?
A 0–5 hours   B 5–10 hours
C 10–15 hours   D 15–20 hours
d) For how many hours was the number of bacteria found to be above 4000 per mm³?
e) Give TWO possible reasons why the number of bacteria began to decrease after 23 hours.

8 An equal volume of fresh, untreated milk was placed in each of four sterile screw-top bottles (W, X, Y and Z). The bottles and their contents were then treated as shown in the table above:
a) The milk failed to turn sour in one of the bottles. Identify it and explain your choice.
b) Predict the order in which the milk in the other bottles turned sour.

9 A type of bacterium called *Escherichia coli* normally lives in the human large intestine. When special culture medium (McConkey's broth) is inoculated with *Escherichia coli* and incubated, the broth changes in appearance from clear red to cloudy yellow.

In the experiment shown in the diagram, samples of water from three rivers X, Y and Z were investigated.
a) Construct a hypothesis concerning the type of pollution affecting river Y to account for the results obtained.
b) Suggest why the test tube is floating in the sample from river Y after incubation.

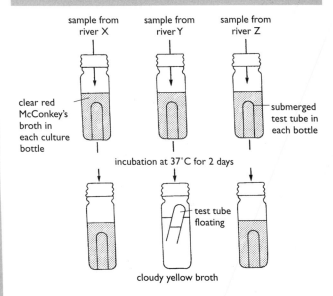

sample from river X   sample from river Y   sample from river Z

clear red McConkey's broth in each culture bottle

submerged test tube in each bottle

incubation at 37°C for 2 days

test tube floating

cloudy yellow broth

25 *Reprogramming microbes*

1 Match the terms in list X with their descriptions in list Y.

list X

1) antibiotic
2) biological detergent
3) genetic engineering
4) immobilisation
5) insulin
6) interferon
7) penicillin
8) plasmid

list Y

a) chemical made by the pancreas that controls blood sugar level

b) first antibiotic to be isolated

c) general name for chemical produced by one micro-organism that prevents growth of some other micro-organisms

d) circular piece of chromosomal material from a bacterium used to transfer a gene from one organism to another

e) cleaning agent containing digestive enzymes produced by bacteria

f) technique by which cells or enzymes are attached to an inert material preventing their free movement and allowing easy reuse

g) transfer of a piece of chromosome from one organism to another producing a reprogrammed organism

h) chemical made by animal cells that prevents multiplication of viruses

2 Study the following shopping list and answer the questions that follow.

quantity	item	price (£)	biotechnological (✓) non-biotechnological (✗)
800 g	bread	0.50	✓
0.5 kg	cheese	1.50	✓
1 kg	apples	1.00	✗
3.5 kg	soap powder	3.00	✓
750 g	yoghurt	1.00	✓
3 kg	potatoes	1.00	✗
1 litre	fruit juice	0.50	✓
2 dozen	eggs	2.50	✗
750 cm^3	wine	3.00	✓
400 g	barbecue sauce	1.00	✓
4	grapefruit	1.00	✗
250 g	butter	0.50	✓
300 cm^3	vinegar	0.50	✓
1 kg	meat	4.50	✗
3 litres	beer	3.50	✓

a) How many biotechnological products are on the shopping list?

b) Express the number of biotechnological to non-biotechnological products as a simple whole number ratio.

c) What percentage of the total amount of cash was spent on biotechnological products?

3 The experiment below was set up to investigate the effect of two antibiotics (penicillin and streptomycin) on the growth of two species of bacteria.

State TWO ways in which the experiment must be altered in order to make it a valid test.

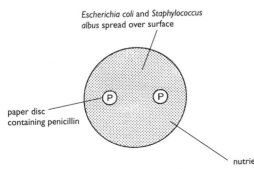

Escherichia coli and *Staphylococcus albus* spread over surface

paper disc containing penicillin

nutrient agar in each plate

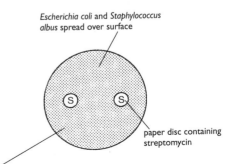

Escherichia coli and *Staphylococcus albus* spread over surface

paper disc containing streptomycin

Problem Solving in Biology

4 A swab was taken from a patient with a throat
   infection and the bacteria were spread over the
   surface of sterile nutrient agar in a petri dish.
       A multidisc with a different antibiotic at the end
   of each of its six arms was then placed on top of
   the bacteria.
       The following diagram shows the result of the
   experiment after 48 hours in a warm incubator.

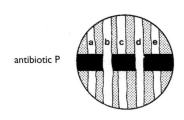

antibiotic P

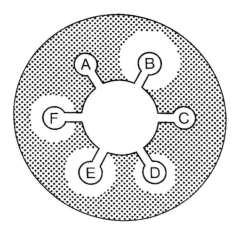

A—F = 6 different antibiotics

[▨▨▨] = zone of bacterial growth

[⬜] = zone of no bacterial growth

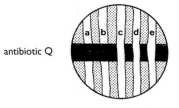

antibiotic Q

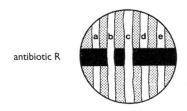

antibiotic R

a) (i) Which antibiotics were able to prevent
growth of the bacterium?
(ii) To which of these was the bacterium most
sensitive? (iii) Explain your answer to part (ii).
b) To how many of the antibiotics was the
bacterium resistant?
c) The patient was known to be allergic to
antibiotics A, B, and F.
(i) Which antibiotic should be used to treat her?
(ii) Give a reason for your choice.

5 Each of the dishes in the experiment shown was
   streaked with five different types of bacteria (a, b, c,
   d and e). A strip containing one of four different
   antibiotics (P, Q, R and S) was then placed across
   each dish.
       The diagram shows the results after incubation
   for 3 days at 37°C.

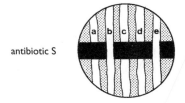

antibiotic S

a) Which antibiotic was least effective at
preventing bacterial growth?
b) Which type of bacterium was sensitive to all
four antibiotics?
c) Which type of bacterium was most resistant to
the antibiotics?
d) Which types of bacteria were sensitive to two
antibiotics and resistant to two antibiotics?

6 Read the following passage.

Lactose is a sugar present in whey (the unwanted remains of milk after the curds have been used for cheese-making). It is composed of two simpler sugars. Brewer's yeast (*Saccharomyces cerevisiae*) is unable to ferment it because the fungus cannot take lactose in through its cell membrane. Even if lactose could enter a yeast cell, the organism lacks the enzyme needed to break the bond holding together the components of lactose – one molecule of glucose and one of galactose.

However, genetic engineers have now solved these problems by transferring pieces of chromosome from a rare yeast (*Kluyveromyces lactis*) into brewer's yeast. The new fungal strain formed possesses the genes required for making both lactose permease (the enzyme that must be present for lactose to enter the cell) and lactase (the enzyme that digests lactose to glucose and

galactose). **As a result normal fermentation can now proceed.**

a) Describe the structure of a molecule of lactose.
b) Give TWO reasons why *Saccharomyces cerevisiae* is unable to use lactose as a foodstuff.
c) With reference only to the information in the passage, describe what is meant by the term genetic engineering.
d) Give TWO reasons why the new strain of yeast formed by genetic engineering is able to make use of lactose.

e) Which of the following chemical reactions is this new microbial strain able to bring about?
A lactase ➡ alcohol   B alcohol ➡ galactose
C glucose ➡ lactose   D glucose ➡ alcohol

7 The four experiments in the following diagram were set up to investigate the action of a new 'biological' soap powder called BIOX. Study them carefully and answer the questions that follow.

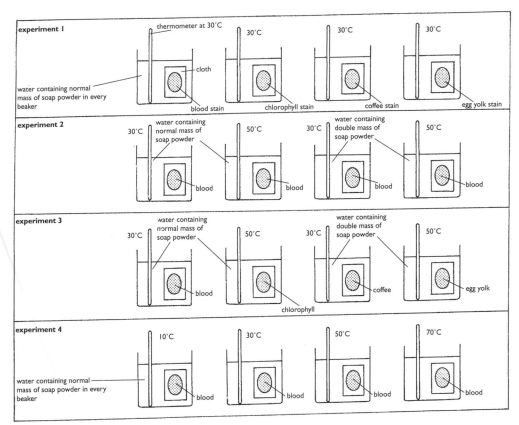

Problem Solving in Biology

a) Which experiment is testing the effect of four different temperatures on the action of BIOX?
b) In which experiment is the type of stain the one variable factor being investigated?

c) Which experiment really consists of two experiments being done at the same time?
d) Which experiment is invalid? Explain why.

8 The following diagram shows one of seven identical fermentation vessels set up to investigate the effect of pellet number on rate of fermentation of glucose by immobilised yeast cells. The results are given in the accompanying table.

number of pellets	rate of fermentation (cm^3 CO_2 released/day)
50	10
100	15
200	27
400	51
500	64
600	64
800	64

a) Draw a line graph of the results.
b) From your graph estimate the fermentation rate that would have been achieved using 300 pellets.
c) What is the optimum pellet number used in this experiment?
d) (i) State the effect on fermentation rate of adding more pellets beyond the optimum number.
(ii) Account for the effect that you gave as your answer to (i).
e) (i) Predict the overall effect on fermentation rate of removing the stirrer from the experimental design.
(ii) Explain your answer to (i).
f) Describe how the experiment could be redesigned to investigate the effect of temperature on rate of fermentation of glucose by immobilised yeast cells.

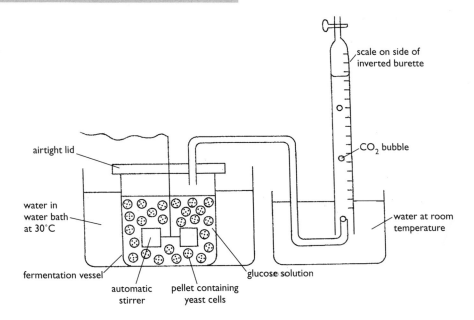

scale on side of inverted burette

CO_2 bubble

airtight lid

water in water bath at 30°C

water at room temperature

fermentation vessel

glucose solution

automatic stirrer

pellet containing yeast cells